DONALD F. BARNETT
ROBERT W. CRANDALL

Up from the Ashes

*The Rise of the Steel Minimill
in the United States*

THE BROOKINGS INSTITUTION
Washington, D.C.

Library of Congress Cataloging-in-Publication data:

Barnett, Donald F.
  Up from the ashes.

  Includes bibliographical references and indexes.
  1. Steel industry and trade—United States.
2. Steel minimills—United States.   I. Crandall, Robert W.
II. Title.
HD9515.B365   1986   338.4′7669142′0973   85-48201

ISBN 0-8157-0834-3
ISBN 0-8157-0833-5 (pbk.)

9 8 7 6 5 4 3 2 1

THE BROOKINGS INSTITUTION is an independent organization devoted to nonpartisan research, education, and publication in economics, government, foreign policy, and the social sciences generally. Its principal purposes are to aid in the development of sound public policies and to promote public understanding of issues of national importance.

The Institution was founded on December 8, 1927, to merge the activities of the Institute for Government Research, founded in 1916, the Institute of Economics, founded in 1922, and the Robert Brookings Graduate School of Economics and Government, founded in 1924.

The Board of Trustees is responsible for the general administration of the Institution, while the immediate direction of the policies, program, and staff is vested in the President, assisted by an advisory committee of the officers and staff. The by-laws of the Institution state: "It is the function of the Trustees to make possible the conduct of scientific research, and publication, under the most favorable conditions, and to safeguard the independence of the research staff in the pursuit of their studies and in the publication of the results of such studies. It is not a part of their function to determine, control, or influence the conduct of particular investigations or the conclusions reached."

The President bears final responsibility for the decision to publish a manuscript as a Brookings book. In reaching his judgment on the competence, accuracy, and objectivity of each study, the President is advised by the director of the appropriate research program and weighs the views of a panel of expert outside readers who report to him in confidence on the quality of the work. Publication of a work signifies that it is deemed a competent treatment worthy of public consideration but does not imply endorsement of conclusions or recommendations.

The Institution maintains its position of neutrality on issues of public policy in order to safeguard the intellectual freedom of the staff. Hence interpretations or conclusions in Brookings publications should be understood to be solely those of the authors and should not be attributed to the Institution, to its trustees, officers, or other staff members, or to the organizations that support its research.

Foreword

In the past decade the U.S. steel industry has lost nearly 20 percent of its capacity. Slow growth in demand for steel, an overvalued dollar, and an overhang of inefficient production facilities are largely to blame. Despite this rather bleak performance, however, the industry has a bright side. A group of small companies called minimills, which melt scrap in electric furnaces, have doubled their capacity in this same period and continue to grow. They have prospered because of their ability to adjust to new technologies, to use plentiful supplies of scrap, and to realize much higher labor productivity than their larger "integrated" rivals.

In this book Donald F. Barnett and Robert W. Crandall document the surprising growth of the minimills, comparing their economic performance with that of the larger steel companies. The authors find that the minimills are increasingly moving into more sophisticated product lines, taking markets away from both importers and the bigger U.S. companies. In fact, the authors show that, unlike their larger domestic rivals, U.S. minimills have costs that are as low as those of any producers in the world. Though the minimills can probably not produce all the higher grades of steel, they will be able to recapture a substantial share of the steel market from imports, thus arresting the rising trend of import penetration.

The success of the minimills, which now number more than forty firms, may provide an example for other industries and for government policymakers. Rather than attempt to protect and rejuvenate large producers, it may be more productive to encourage new, entrepreneurial firms to grow. The minimills provide at least one example of a domestic enterprise that has risen virtually from the ashes of a seriously troubled industry.

Donald F. Barnett, formerly chief economist and vice president of the American Iron and Steel Institute, is an independent consultant and a member of the faculty of the University of Windsor. Robert W. Crandall is a senior fellow in the Brookings Economic Studies program. The authors are particularly grateful to Alice M. Rivlin for her helpful

vii

suggestions, and to Edward Flom, Kenneth Iverson, John Tumazos, and Joseph Wyman for their valuable comments on an early draft of the manuscript. They are also grateful to Menzie Chinn and Elizabeth A. Schneirov for research assistance, to Dee Koutris and David Rossetti for typing the manuscript, to Alice M. Carroll and Nancy Davidson for editing it, and to Carolyn A. Rutsch and Almaz Zelleke for checking its factual content.

The views expressed here are those of the authors and should not be ascribed to the persons whose assistance is acknowledged, or to the trustees, officers, or other staff members of the Brookings Institution.

BRUCE K. MACLAURY
President

August 1986
Washington, D.C.

Contents

Appendixes

Text Tables

Appendix Tables

Figures

Up from the Ashes

I

Two Distinct Industries

There is a widespread belief that the United States is losing its basic industry, that services are replacing goods production, and that capital is flowing only into new, high-technology industries. Surprisingly, there has been growth in one of the most basic of U.S. industries—steel. Even as steel consumption has declined around the world and numerous producers have failed, a new breed of steel producer—the minimill—has emerged. These minimills have grown rapidly in number and in size in the United States in recent years, more than doubling their output from 1975 to 1985.

While the minimills have been growing, the larger U.S. steel companies have been reducing their production capacity. Rising costs in the face of declining prices and output levels caused almost a one-third reduction in Big Steel's capacity between 1975 and 1985, and much of the remaining production capacity may not be viable. Several large producers, such as LTV, McLouth, Kaiser, and Wheeling-Pittsburgh, have disappeared or have been reorganized by bankruptcy courts. And several others are in severe financial difficulty. Yet the newer minimill companies, such as Nucor Corporation, North Star Steel Company, and Florida Steel Corporation, have been adding capacity.

Once a highly concentrated industry made up of a few integrated plants capable of producing millions of tons annually, steel is being transformed by a dynamic group of young firms producing between 200,000 and 1 million tons per year. These small firms, with narrow product lines, are usually far more efficient than most of their larger rivals. When challenged, the larger firms often abandon product lines to their competitors rather than do battle with the upstarts. Obviously, the relative costs of the two types of firms are a major factor in this metamorphosis.

This book describes the remarkable growth of the minimills and attempts to analyze the sources of their competitive advantage over Big Steel. Factor prices, innovation, and product quality are important elements in the ongoing invasion of an old industry by a new breed of

1

steel entrepreneurs. This study seeks to explain why the Davids continue to grow in strength and number while the Goliaths quietly decline. It also attempts to determine whether there are definable limits to the growth of the minimills.

The rise of the small steel companies provides insights into the changes that are gripping many parts of the American economy. The U.S. automobile industry, for instance, is facing the end of the trade protection it has enjoyed from Japanese imports since 1981. Textiles, motorcycles, copper, machine tools, and many other heavy basic industries have demanded protection, arguing that an overvalued dollar makes it impossible for them to compete with foreigners. The small steel companies, however, have thrived with little trade protection. Indeed their costs have often been substantially lower than those of their international rivals. Aggressive, innovative management and low prices for the scrap metal they use as a raw material have combined to produce a highly competitive industry within an industry. In short, the steel industry is evidence of how Schumpeter's "creative destruction" of monopoly power works. Will other basic industries travel the same path?

The chapters that follow attempt to explain the success of the minimills in economic terms. They compare production costs of the larger and smaller firms; analyze the course of two vital production requirements, scrap and electricity supply; attempt to project future technological changes; and question why a strategy undertaken by the new entrepreneurs is unavailable to Big Steel. The concluding chapter suggests that policymakers should pay at least as much attention to the newer companies' strategies as to the larger companies' dilemmas in shaping U.S. policy toward the steel industry.

Changing Production Processes

To most people, steel production means giant facilities engaged in the transformation of iron ore, coal, and limestone into molten metal. Companies producing steel are thought to be large because of the huge scale of their facilities. In fact, many steel companies are quite small, with capital of less than $100 million. In the United States alone, there are nearly fifty of these companies, known as minimills, and the number is growing. This minimill sector has recently risen up to take a lion's share of the market from its larger brethren, the integrated firms.

Among the biggest investments in the steel-making industry, and the reason for the very large scale of production among the steel giants, is the blast furnace. The integrated production process that characterizes the large firms depends on blast furnaces to combine iron ore, coal, and limestone to produce the pig iron, or "hot metal," that is then converted into molten steel (see figure 1-1). Improvements in blast furnace technology over the past three decades have caused the minimum efficient scale of these furnaces to increase substantially. The Japanese, for example, have developed methods of combining the basic raw materials in very large vessels, capable of producing more than 3 million tons of hot metal per year. While the efficiency gains from furnaces with a capacity of more than 1.5 million tons may not be great, even a 1.5 million-ton blast furnace translates into a large steel plant. Because efficient design and flexible operation must allow for periodic relining of the furnace's interior walls, an integrated operation generally requires at least two blast furnaces. This means that the minimum efficient scale for integrated steel making is at least 3 million tons of molten pig iron annually. In Japan, the average plant has more than 7 million tons of capacity.[1]

The basic oxygen furnace (BOF), which converts the hot metal from the blast furnace into molten steel, has made the traditional open hearth furnace virtually obsolete. Scrap is also charged, or loaded, into basic oxygen furnaces and can account for as much as 33 percent of the metallics (and even more if the scrap is preheated). By the end of the 1980s, all U.S. integrated works probably will use the BOF technology because it requires a far lower refining ("heat") time than the open hearth. Some integrated plants (such as Bethlehem Steel Corporation's Johnstown plant) have closed their blast furnaces and coke ovens and replaced their open hearths with electric steel-making furnaces to feed their older rolling facilities. However, the electric-furnace capacity in the integrated firms is rarely part of an optimally designed, small-scale facility.[2]

Basic oxygen furnaces, because of their short heat time, have a rather

1. Richard Serjeantson, Raymond Cordero, and Henry Cooke, eds., *Iron and Steel Works of the World*, 8th ed. (Surrey, England: Metal Bulletin Books, 1983), pp. 303–45.

2. Among the exceptions to this rule are a large plant in Baytown, Texas, with a plate mill that supplies United States Steel's welded pipe mills, and an Armco plant in Kansas City, which is the only integrated-company plant that even resembles a minimill.

Figure 1-1. *Steel Production Processes*

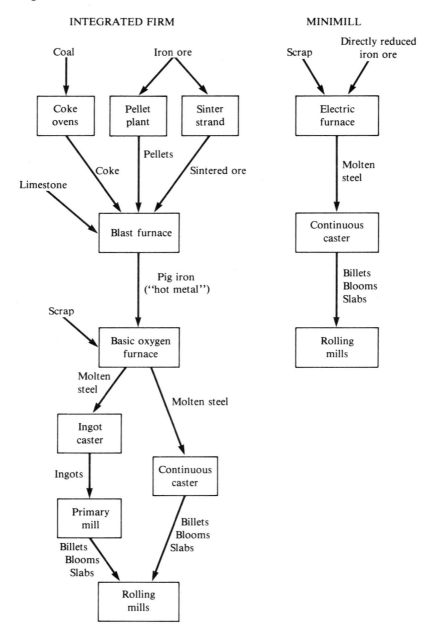

large minimum efficient scale. Most BOF facilities in the United States have been built with two or three vessels, each with 200 to 300 tons of capacity. This translates into an annual production capacity of 3 million to 5 million tons of molten steel, which allows time for routine maintenance and scheduled relining.

Molten steel may either be poured into ingots, which then must be sent on to primary mills for rolling into semifinished shapes, or be poured directly into a continuous caster. The continuous caster, developed in the past two decades, forms molten steel directly into semifinished shapes—the slabs, blooms, and billets fed into rolling mills that turn out the industry's final product. The continuous caster not only bypasses the reheating required before the primary rolling of ingots but produces steel with more consistent metallurgical properties than the ingot process.

Integrated companies in the United States have been much slower to adopt continuous casting than European or Japanese companies. Casters are difficult to fit into existing plants, and U.S. integrated firms have not built any new plants in the last twenty years. Large casters for producing slabs are very expensive to install, requiring outlays of between $150 million and $250 million. By the time that rising energy costs had made clear the economic advantages of replacing ingot lines and primary mills with slab casters, the large U.S. companies were finding it difficult to make large capital outlays. The major impetus to their installation of continuous casting came in the late 1970s and 1980s in response to customers' demands for quality.

The integrated plants continue to be the principal source of production for the steel industry's largest semifinished shape, the slab. Slabs, which are generally eight to twelve inches thick and several feet wide, are rolled into sheet products that are used in automobiles, appliances, containers, and a variety of other consumer and capital goods. For efficient operation, the continuous hot-strip mills used in the final step of the integrated production process must roll about 3 million tons of slabs annually. These high-speed mills, capable of rolling slabs to only a few millimeters of thickness, may cost $600 million or more; they therefore are not consonant with minimill operations.

Minimills typically produce only billets, the smallest semifinished shapes, which are suitable for rolling into bars, small structural shapes, and rods. These mills simply charge scrap into an electric furnace to produce molten steel. With only a few exceptions, they use continuous casting for all of their output.

Most minimills are of very recent vintage, employing the latest technology in furnaces, continuous casters, and rolling mills. Indeed, it is not unusual for a minimill to refurbish or completely replace a production unit after only a few years of use. Unlike the integrated companies, whose facilities have an average age of about twenty years, the minimills construct their facilities anticipating that they will shortly be outmoded.[3]

The Integrated Sector

In the United States, there are fourteen companies that can be defined as integrated steel firms. These range from the United States Steel Corporation, a large steel and oil company, to California Steel and Tube, which acquired the assets of Kaiser Steel Corporation's Fontana plant and is operating only the rolling mills of this plant. Together, the fourteen had a total capacity in 1985 for producing 105.6 million tons of raw steel annually. Their individual capacities were:[4]

Firm	Raw steel capacity (millions of tons per year)
United States Steel Corp.	26.2
LTV Steel Corp.	19.1
Bethlehem Steel Corp.	18.0
Inland Steel Co.	9.3
Armco	6.8
National Steel Corp.	5.6
Wheeling-Pittsburgh Steel Corp.	4.5
Weirton Steel	4.0
Ford Motor Co. (Rouge Steel Co.)	3.6
California Steel	2.1
McLouth Steel Products Corp.	2.0
CF&I Steel Corp.	2.0
Interlake	1.4
Sharon Steel Corp.	1.0

3. See D. F. Barnett, "The American Steel Industry in the 1980's—Capital Requirements for Modernization," in *Economic Papers on the American Steel Industry* (Washington, D.C.: American Iron and Steel Institute, 1981), p. 13.

4. Estimated from data in company reports; Oppenheimer and Co. periodic metal industry reports; Serjeantson, Cordero, and Cooke, eds., *Iron and Steel Works of the World*. CF&I closed its integrated steel-making facilities in 1983, and California Steel's raw-steel facilities (bought from Kaiser) remained closed.

Table 1-1. *Share of U.S. Raw Steel Produced in Electric Furnaces,*
1970–85

Year	Raw steel production (millions of tons)		Electric furnace as a percent of total production
	Electric furnace	Total	
1970	20.2	131.5	15.3
1971	20.9	120.4	17.4
1972	23.7	133.2	17.8
1973	27.8	150.8	18.4
1974	28.7	145.7	19.7
1975	22.7	116.6	19.4
1976	24.6	128.0	19.2
1977	27.9	125.3	22.2
1978	32.2	137.0	23.5
1979	33.9	136.3	24.9
1980	31.2	111.8	27.9
1981	34.1	120.8	28.3
1982	23.2	74.6	31.1
1983	26.6	84.6	31.5
1984	31.4	92.5	33.9
1985	29.9	88.3	33.9

Source: American Iron and Steel Institute, *Annual Statistical Report, 1985* (Washington, D.C.: AISI, 1986), table 25, and earlier issues. Figures are rounded. Tons in all tables are net tons unless otherwise noted.

Each of these firms has suffered substantial losses in the 1980s. Most now serve a smaller share of the market than they formerly had and have retired some of their production capacity. Three have suffered bankruptcy (LTV Corporation, McLouth Steel Products Corporation, and Wheeling-Pittsburgh Steel Corporation) and one has closed—Kaiser (forerunner of California Steel); one is now owned by its employees (Weirton); and several are the product of mergers designed to rationalize industry assets.

The Minimill Sector

The minimills' share of raw steel production is difficult to measure because industry statistics do not distinguish them from the integrated producers. Raw steel production has fallen precipitously since the 1970s, but the proportion of the total coming from electric furnaces has increased steadily, reaching about 34 percent in 1985 (see table 1-1). Most of the

Table 1-2. *Raw Steel Capacity and Products of Minimill Companies in the United States, 1986*

Company[a]	Number of plants	Annual capacity (thousands of tons)	Products[b]
North Star Steel Co.	5	2,500	Bars, wire rods, small shapes, pipe and tube
Nucor Corp.	4	2,100	Bars, small shapes
Northwestern Steel and Wire Co.	1	1,800	Bars, wire rods, small and large shapes
Florida Steel Corp.	5	1,580	Bars, wire rods, small shapes
Chaparral Steel Co.	1	1,100	Bars, small and large shapes
Structural Metals	2	800	Bars, small shapes
Atlantic Steel Co.	2	750	Bars, wire rods, small shapes
Raritan River Steel Co.	1	750	Wire rods
Laclede Steel Co.	1	700	Bars, wire rods, large shapes
Bayou Steel Corp.	1	650	Bars, small and large shapes
Georgetown Steel Corp.	1	600	Bars, wire rods, small shapes
Continental Steel Corp.	1	600	Bars, wire rods
Keystone Steel & Wire Co.	1	600	Wire rods
Newport Steel	1	550	Pipe and tube
Roanoke Electric Steel Corp.	1	480	Bars, small shapes
Birmingham Steel	3	460	Bars
Quanex Corp.	2	440	Bars, pipe and tube
Seattle Steel	1	400	Bars, small shapes, plates
Tamco	1	400	Bars, wire rods
Sheffield Steel Co.	1	350	Bars
Oregon Steel Mills	1	350	Plates
Thomas Steel Corp.	1	300	Bars, small shapes
Steel of West Virginia	1	270	Small shapes, plates
Cascade Steel Rolling Mills	1	250	Bars, small shapes
New Jersey Steel Corp.	1	250	Bars, small shapes
Auburn Steel Co.	1	240	Bars, small shapes
Knoxville Iron Co.	1	225	Bars
Northwest Steel Rolling Mills	1	220	Bars, small shapes
Razorback Steel Corp.	1	220	Bars, small shapes, railroad tie plates
Kentucky Electric Steel Co.	1	200	Bars, small shapes
Marion Steel	1	200	Bars
Milton Manufacturing	1	200	Bars
Border Steel Mills	1	200	Bars
Judson Steel Co.	1	190	Bars

Table 1-2 *(continued)*

Company[a]	Number of plants	Annual capacity (thousands of tons)	Products[b]
Roblin Steel	1	180	Bars
Calumet Steel Co.	1	150	Bars, small shapes
Charter Electric Melting	1	130	Bars
Intercoastal Steel Corp.	1	100	Bars, wire rods
Owen Electric Steel Co. of South Carolina	1	100	Bars
Hurricane Industries	1	100	Bars
Marathon LeTourneau Co.	1	75	Plates
Hawaiian Western Steel	1	60	Bars

Sources: Richard Serjeantson, Raymond Cordero, and Henry Cooke, eds., *Iron and Steel Works of the World,* 8th ed. (Surrey, England: Metal Bulletin Books, 1983); *33 Metal Producing: Mini-mill Handbook* (McGraw-Hill, 1983); *33 Metal Producing: Steel Industry Data Handbook—USA 1985* (McGraw-Hill, 1985); and telephone interviews with company sources.

a. Ranked by size of production capacity.

b. Bars include reinforcing bars.

increase has come from minimills, which accounted for about 20 percent of all raw steel production in the United States in 1985.

Similarly, about 16.5 percent of all U.S. steel-making capacity is in the minimill sector of the industry. The striking thing about this sector is the number of firms (sixteen) with a capacity of 200,000–400,000 tons per year of raw steel (see table 1-2). The fact that such small plants can survive and even prosper suggests that the minimum efficient scale for makers of basic bar products is extremely small. Almost two-thirds of all minimill capacity is in plants with less than 600,000 tons of crude steel capacity per year (table 1-3).

Most minimills specialize in standard bar products, small structural shapes, or wire rods rolled from small-diameter billets (six inches or less in cross section).[5] These products typically are made from scrap that contains impurities that are difficult to eliminate without costly presorting or ladle metallurgy. Some producers, such as Northwestern Steel and Wire Company and Raritan River Steel Company, however, have succeeded in producing wire rods of high quality. And others are producing bars of high quality (including the feedstock for cold-finished

5. These products are typically made from carbon steel, not alloy or stainless steel, though minimills have begun moving into the higher grades of steel.

Table 1-3. *Distribution of Minimill Plants, by Production Capacity, 1985*

Capacity range (thousands of tons per plant)	Number of plants	Total capacity (thousands of tons)	Percent of total
1,200 and over	1	1,800	8.3
1,000 to 1,200	1	1,100	5.0
800 to 999	1	900	4.1
600 to 799	6	3,900	17.9
400 to 599	15	7,030	32.2
200 to 399	22	5,585	25.6
Up to 200	12	1,505	6.9
Total	58	21,820	. . .

Source: See table 1-2.

bars) and alloy bars. "Special quality" bars are, in fact, becoming commonplace in the minimill sector.

Minimills are expanding their production into the domain of integrated producers with great regularity. Until recently, integrated firms dominated the production of pipes and tubes. Copperweld Steel Company, a foreign-owned producer, and Ipsco (Interprovincial Steel and Pipe Corporation), a Canadian firm, were the only small firms producing tubular products until the new minimill entrants began appearing. Bayou Steel Corporation and Chaparral Steel Company have moved into larger structural shapes, and both are shipping their products considerable distances. Table 1-4 indicates the range in products of the minimills and integrated firms.

With recent technological developments in the casting of thin slabs and the simplification of rolling techniques, minimills in the United States should be capable of producing sheet products by the late 1980s. A few minimills are being built in other countries to produce sheet products using conventional slab casters. A mill under construction in Tuscaloosa, Alabama, promises to be a very low cost producer (see chapter 4). It is likely that many minimills will be designed for flat-rolled products (sheets and plates) in the United States.

Prosperity and Decline within a Single Industry

Since 1974, the world steel industry has languished. Slow economic growth and the use of lighter materials, along with rising energy prices,

Table 1-4. *Major Steel Products of Integrated Firms and Minimills, 1985*

Semifinished shape and product	Produced by	
	Integrated firms	Minimills
Slabs[a]		
Hot-rolled sheets	x	. . .
Cold-rolled sheets	x	. . .
Coated sheets	x	. . .
Plates	x	x
Welded pipe and tube	x	x[b]
Blooms and billets[c]		
Wire rods	x	x
Bars	x	x
Reinforcing bars	x[b]	x
Small structural shapes[d]	x[b]	x
Large structural shapes[e]	x	x[f]
Rails	x	. . .
Seamless tubing	x	x
Axles and wheels	x	. . .

Source: Serjeantson and others, eds., *Iron and Steel Works*, pp. 693–818.
a. Slabs are the largest semifinished shapes, generally 8–12 inches in cross section and several feet wide.
b. Limited production.
c. Blooms are square semifinished shapes, 6–8 inches in cross section; billets are also square and less than 6 inches in cross section.
d. Less than 3 inches in cross section.
e. Three inches or more in cross section.
f. Limited production above 12 inches in cross section.

have kept world steel production below 1974 levels.[6] Production in the United States and other developed countries has been particularly sluggish because of a continual shift of steel consumption and production toward less developed economies. Raw steel output in the United States peaked at 150.8 million tons in 1973, fell to an average of about 130 million tons annually in the late 1970s, and collapsed in the 1980s in the wake of a rising dollar (which stimulated imports) and recession. Production fell to a dismal 74.6 million tons in 1982, and in 1985 it had rebounded marginally to 88.3 million tons (see table 1-1).

Despite this sharp decline in the industry's output, minimills have steadily expanded their capacity and output. Since the early 1970s, they have more than doubled their raw-steel capacity (table 1-5), while total U.S. steel consumption has declined by about 12 percent. Their output

6. International Iron and Steel Institute, *Steel Demand Forecasting* (Brussels: IISI, 1983).

Table 1-5. *Raw Steel Production in the United States, Various Years, 1960–85*
Millions of tons

Year	Minimills		Integrated and specialty firms		Total	
	Capacity	Output	Capacity	Output	Capacity	Output
1960	2.8	2.0	140.0	97.3	142.8	99.3
1965	4.5	3.7	143.7	127.8	148.2	131.5
1970	7.5	7.0	146.3	124.5	153.8	131.5
1975	10.2	7.8	142.9	108.8	153.1	116.6
1980	15.5	13.5	138.2	98.3	153.7	111.8
1985	21.8	17.6	111.8	70.7	133.6	88.3

Source: Authors' estimates based on data from producers and AISI, *Annual Statistical Report, 1985,* tables 1A and 1B, and earlier issues. Output figures for 1985 are projections.

rose from 7 million tons of raw steel in 1970 to almost 18 million tons in 1985. The raw steel output of integrated and specialty producers, on the other hand, reached 124.5 million tons in 1970 but by 1985 was just 70.7 million tons.[7]

In 1975, the American Iron and Steel Institute was predicting recurrent steel shortages if the industry did not expand its capacity by 30 million tons by 1980.[8] Most of the larger companies invested rather heavily in raw materials and had plans to expand their facilities, anticipating a sharp rebound from the 1975 recession. The recovery never came, and the integrated producers were forced to begin reducing capacity as early as 1977.

The difficulties of U.S. producers in the 1980s have been exacerbated by excess world steel-making capacity and the strong dollar. Steel companies in Europe, Japan, and less developed countries had also prepared for growing demand. "Greenfield" projects—new plants, on new sites—had been launched in France, West Germany, South Korea, Brazil, Mexico, Japan, and a number of other countries in the 1970s. Though the failure of recovery in 1976 had interrupted these projects, many were at least partially completed, and other facilities were expanded and modernized.

7. "Specialty steel" is high-quality alloy and stainless steel. Most of this steel is produced by a group of small companies using very high quality scrap in electric furnaces and a number of sophisticated metallurgical technologies. These are not typically referred to as minimills.

8. American Iron and Steel Institute, "Financing Capital Expenditures: A Critical Concern of the American Steel Industry," statement prepared for the Council on Wage and Price Stability, October 30, 1975, table A-2.

The governments of less developed countries and of countries in the European Community often provided the capital and operating funds to allow facilities that would have closed or been reorganized to continue in production. Government ownership in less developed countries— Mexico and Brazil, for example—accelerated the shift of steel production to the third world. Though the shift would probably have occurred anyway because of lower labor and construction costs, government ownership allowed production to expand without regard to the immediate economics of the investment decision.

Many of the producers in Europe and developing countries were given added impetus to produce for export to the United States by the strong dollar. Between 1980 and early 1985, the dollar appreciated two-thirds against a Federal Reserve trade-weighted average of foreign currencies in real terms.[9]

Imports and the U.S. Market

It is commonplace to ascribe the troubles of integrated steel producers in the United States to rising imports, but this is a decided oversimplification. Until the 1980s, imports generally remained in a range of 15 percent to 18 percent of domestic consumption.[10]

U.S. steel consumption fell substantially between 1974 and 1985. It plunged from 120 million tons in 1974 to a nadir of 76 million tons in 1982 and had rebounded to only 96 million tons by 1985, the third year of a strong economic recovery. During the same period, imports rose from 16 million tons to 24 million tons as the result of a strong dollar in 1985, and exports fell from 6 million tons to 1 million tons. Thus the market for all U.S. steel producers dropped from 110 million tons in 1974 to 73 million tons in 1985.[11] During this time, minimills increased their shipments from about 8 million tons to 15 million tons. Therefore the share of the market for the large integrated producers and specialty steel firms fell from 102 million tons to 58 million tons in just eleven years. Of the 44-million-ton decline in the integrated producers' market, only 8 million tons may be attributed to imports and 7 million tons to the expansion of

9. *Economic Report of the President, February 1986*, p. 373.

10. AISI, *Annual Statistical Report, 1984* (Washington, D.C.: AISI, 1985), tables 21 and 27, and earlier issues.

11. Ibid., *1985; 1982*, tables 1A and 1B.

the U.S. minimills. Of the balance, 24 million tons is due to the decline in U.S. consumption and 5 million tons to falling exports.

The combination of import competition, minimills' penetration of the market, and the sharp decline in steel consumption has been devastating to the integrated producers. As table 1-5 shows, they reduced their production capacity by approximately 35 million tons between 1970 and 1985, but they still had excess capacity that was not profitable at 1985 prices.

The strong dollar made the U.S. market in 1985 among the highest-priced markets in the developed world, thereby attracting imports. The competition from imports and minimills kept domestic prices virtually constant between 1981 and 1985. Some minimills, such as Nucor, were able to operate at a profit in those years, despite this weakness in prices. The integrated companies, on the other hand, lost an average of about $100 per ton in 1982, $50 per ton in 1983, $9 per ton in 1984, and $11 per ton in 1985.[12]

From 1976 through 1985, the average integrated producer's common stock price fell by nearly 50 percent while the average New York Stock Exchange share rose by about 80 percent (table 1-6). During this same period, the common share prices of the two major publicly held minimill firms—Florida Steel and Nucor—rose by 164 percent and 647 percent. Quanex's stock underperformed the market because it specializes in tubular products, whose demand collapsed in 1982. Texas Industries, which owns Chaparral,[13] enjoyed a substantial rise in its share price, but this was undoubtedly due in large part to its other construction products. Keystone Consolidated Industries is a diversified company, but its performance was affected by the relatively poor performance of Keystone Steel and Wire Company, a rather old minimill. Northwestern Steel and Wire, with three enormous 400-ton furnaces that were not designed to mesh with continuous casters, is not a minimill in the modern sense. It encountered three years of large operating losses in 1982–84 as it attempted to restructure its operations. It is difficult to assess the performance of the other major minimill companies because they are either part of foreign companies (Bayou, Auburn Steel Company, and Raritan River), or of diversified U.S. firms (North Star), or are simply privately held.

12. John Tumazos, "Metals Industry Action Notes" (New York: Oppenheimer and Co., February 5, 1986).

13. In November 1985, Texas Industries acquired the 50 percent of Chaparral Steel Co. it had not previously owned from Co-Steel International of Canada.

Table 1-6. *Performance of Minimill and Integrated Steel Companies on the New York Stock Exchange, 1976–85*

Sector and company	Price range of common stock (dollars)		Percent change, 1976–85	Average market-to-book ratio, 1985[a]
	1976	1985		
Minimills				
Florida Steel Corp.	4.88–7.44	12.63–19.88	163.88	1.08
Keystone Consolidated Industries	16.38–22.13	2.50–5.00	−80.53	0.17
Northwestern Steel and Wire Co.	27.50–36.13	8.00–14.50	−64.64	0.44
Nucor Corp.	3.70–7.91	31.00–55.75	647.20	2.05
Quanex Corp.	5.00–7.41	5.00–10.50	24.90	0.95
Texas Industries (owner of Chaparral Steel Co.)	11.00–15.38	25.75–34.38	127.94	1.33
Integrated firms				
Armco	17.83–23.67	6.75–11.63	−55.71	0.67
Bethlehem Steel Corp.	33.00–48.00	12.50–21.13	−58.52	0.84
Inland Steel Co.	41.00–58.63	19.50–26.00	−54.32	0.54
LTV Steel Co.	10.00–17.75	5.25–13.25	−33.45	0.72
National Steel Corp.	37.38–52.25	24.00–33.63	−35.71	0.64
Republic Steel Corp.[b]	27.13–40.38	15.12–20.50	−47.25	. . .
United States Steel Corp.	45.68–57.50	24.38–33.00	−44.39	0.63
Wheeling-Pittsburgh Steel Corp.	16.00–23.88	6.38–18.13	−38.54	0.58
N.Y. Stock Exchange index Average monthly closing price	60.40	109.20	80.79	. . .

Sources: *New York Times*, January 1, 1986, and January 1, 1977; Standard and Poor's Corp., *Standard Corporation Descriptions, 1986* (McGraw-Hill, 1986).

a. Average of high and low equity price divided by book value, end of fiscal 1984.

b. Merged into LTV Steel Co. in 1984; price data run through 1983.

The difference between the minimills and the integrated firms is also illustrated by recent import trends. The integrated steel companies are the only important domestic suppliers of sheet products, large structural shapes, and most plate categories. Sheet imports were 16–19 percent of domestic shipments between 1981 and 1984 as the integrated firms kept their prices within 5 percent of the delivered prices of imports (table 1-7). For the heavier structural shapes and plates, however, imports have soared because domestic prices have been far above the delivered prices of imports.

For the smaller products—wire rods and bars—domestic prices have fallen relative to import prices, and import shares have actually declined since 1970 despite the strong dollar in the early 1980s. This obviously is

Table 1-7. *Ratios of Domestic to Import Prices and of Imports to Domestic Shipments in U.S. Steel Market, Various Years, 1971–84*

Sector and product	Ratio of domestic price to delivered import price			Ratio of imports to domestic shipments		
	1971–75	1976–80	1981–84	1971–75	1976–80	1981–84
Integrated firms						
Hot-rolled sheets	0.91	1.04	1.04	0.14	0.14	0.16
Cold-rolled sheets	0.94	1.05	1.06	0.19	0.17	0.19
Plates	1.02	1.24	1.30	0.23	0.35	0.53
Structural shapes	1.00	1.17	1.12	0.25	0.42	0.60
Minimills						
Hot-rolled bars	1.02	1.11	1.00	0.10	0.08	0.08
Wire rods	1.10	1.07	0.98	0.72	0.38	0.37

Sources: U.S. Bureau of the Census, *U.S. Imports for Consumption and General Imports, TSUSA Commodity by Country*, FT 246, various years; Bureau of the Census, *Current Industrial Reports: Steel Mill Products*, MA-33B, various years.

the result of intense competition among domestic minimills, whose numbers have been growing rapidly since 1970. Note the sharp decline in the relative domestic price and the import share for wire rods in the late 1970s and 1980s after the entry of Georgetown Steel Corporation and Raritan River.[14]

Summary

It may seem ironic that so many small steel companies are flourishing in a declining U.S. steel market that has witnessed the closing of more than 35 million tons of capacity. Minimills have made sizable inroads into the market and now account for 20 percent of the U.S. steel industry's output. These small companies have provided an alternative to traditional integrated production in a number of lines and have offered the first major challenge to imports.

Clearly, minimills must enjoy substantial advantages over their larger, integrated brethren. Otherwise, they could not have expanded rapidly in a declining market. Among the most important advantages are their higher productivity, lower wages, product specialization, and geographic specialization, their use of cheap scrap, and their lower capital costs. How minimills attained these advantages is the subject of the next chapter. Chapter 3 explores the problems of the integrated firms in competing with the minimills, and chapter 4 examines the prospects of

14. For further discussion of recent price behavior, see chap. 2.

future minimill assaults on established markets of the integrated sector. The principal elements in the operating costs of minimills—scrap and electricity—are examined in some detail in chapter 5. The concluding chapter estimates the equilibrium shares of minimills and integrated companies by the end of the century.

II

The Competitive Position of Minimills

In an industry in which large firms have traditionally been dominant, companies producing only 200,000 to 2 million tons of steel per year would seem to be fringe firms living under the giants' pricing umbrella. In fact, like the new airlines during the first few years of deregulation, the new, smaller firms in the steel industry have provided the competitive spur, cutting prices and increasing their share of the market while many of the larger firms have been trapped in their older, high-cost methods of doing business.[1]

A Historical Perspective

The origin of today's minimills can be traced to the 1930s, when Northwestern Steel and Wire Company began using an electric furnace to produce carbon steel. Up to that point the electric furnace had been used primarily for refining steel and producing specialty (carbon and alloy) steel that required slow heating. Northwestern pioneered in the use of larger and larger electric furnaces and, with other small firms, brought about dramatic improvements in their technology. The length of time needed to heat the metallic charge was shortened, quality control strengthened, electrode and power usage improved, and the life of the refractories that line the furnaces extended. Indeed, the large electric furnaces proved so efficient that some of the large steel companies installed them in their integrated plants to produce carbon steel from scrap.

The rise of the minimill sector in the United States began in the 1960s with Florida Steel Corporation's installation of small electric furnaces to produce construction-grade steel from scrap. Initially the company had a small-scale facility, with a capacity of less than 100,000 tons, for melting and rolling concrete reinforcing bars. Since that time, numerous

---

1. See John R. Meyer and Clinton V. Oster, Jr., *Deregulation and the New Airline Entrepreneurs* (MIT Press, 1984), chap. 7.

similar small-scale mills have been built and rapid improvements have been made in electric furnace technology. Introduction of continuous casting brought dramatic decreases in costs, and further economies have been realized because of reduced power use, improved heat times, and substitution of water-cooled panels for refractories, as well as other changes in refractory design.

By the end of the 1960s, the minimills were supplying about 5 percent of the U.S. steel market, producing concrete reinforcing bars, light shapes used in construction, and merchant bars for use in manufacturing. Up to this point the minimills pursued a regional strategy, locating near markets and scrap supplies, and away from their integrated competitors, protected from them by the high transportation costs that limit the geographic scope of steel markets.

In the early 1970s, however, minimills began to be competitive with integrated mills. As new minimills were built, each more efficient and lower in cost than the last, they rapidly displaced integrated mills in the production of small-diameter products. With an explosive growth, from about 5 percent to 12 percent of raw steel production in the 1970s, they diversified into more and more sophisticated products—wire rods, higher quality bars, and medium-sized structural shapes. The mills became ever larger, but each mill continued to have a very narrow specialization. The combination of product specialization and simplified techniques of production made it possible for minimills to realize economies of scale at a lower level of output than most integrated plants.

By the late 1970s, "market" minimills had emerged, not limiting their sales to a region but producing a very narrow range of products, very efficiently, for a broad geographic market area. Simultaneously, minimills began to adhere to the concept of a single, medium-sized electric furnace balanced with one continuous caster and one rolling mill to give maximum efficiency (rather than using either a large furnace to get maximum scale or multiple small furnaces for operational flexibility).

The minimills have steadily gained on the integrated mills in terms of competitiveness, capturing more and more of each market they have penetrated, and moving on to the more sophisticated products formerly the province of the integrated companies. The minimills have not only pushed aside integrated U.S. producers but have also captured a larger share of the market once held by imports. Were it not for the soaring value of the U.S. dollar in the early 1980s, the minimills would have displaced even more imports.

Minimill entry has been easy, especially in the lower product grades, because of the low capital costs of minimills' plants. But this ease of entry has also created too much capacity in lower-grade products, forcing some plants to close in 1985 and threatening others.

Ingredients of Growth

Why was the growth of minimills so rapid? The failure of their competitors reveals none of the reasons for their success. Among the more obvious advantages that the minimills enjoy over their larger integrated rivals are higher productivity, lower wages, product specialization, geographic specialization, low-cost scrap, low capital costs, and a sound overall strategy for building on strengths and overcoming weaknesses.

Superior Productivity

Labor productivity (man-hours required to produce a ton of steel) in minimills is much higher than in integrated mills in the same product lines. For example, the number of man-hours per ton required to produce wire rods in a typical U.S. minimill is half of that in a typical U.S. integrated mill and 60 percent of that in a Japanese integrated mill (table 2-1). In part, this advantage reflects the use of electric furnaces instead of a combination of coke ovens, blast furnaces, and basic oxygen furnaces. However, even in the same processes, the minimills' productivity is superior. Most minimills use the newest technology, but they also tend to operate more efficiently in any given process—for example, tap-to-tap times, between pourings of heats or batches of molten steel, are much shorter and man-hours per ton much lower in minimill electric furnaces. This was not always the case. In the 1960s, the minimills' performance was relatively poor, but they have done far more to improve productivity and other efficiency measures than the integrated companies.

Their success can be traced to their rapid implementation of new labor-saving technologies and to their efforts to improve the operation of their facilities. From their inception, minimills were built on the assumption of a short economic life, and they have been replaced or modernized rapidly. Each new generation of minimills has embodied the latest techniques in furnaces, casters, and rolling mills. And constant

Table 2-1. *Cost of Producing Wire Rod in Representative U.S.*
Minimills and U.S. and Japanese Integrated Firms, 1981 and 1985

Item	U.S. integrated firms 1981	U.S. integrated firms 1985	U.S. minimills 1981	U.S. minimills 1985	Japanese integrated firms 1981	Japanese integrated firms 1985
	Dollars per ton of finished product					
Operating costs	363.00	339.00	262.00	244.00	265.00	257.00
Labor	127.00	112.00	59.00	42.00	51.00	47.00
Iron ore	61.00	61.00	. . .	. . .	48.00	44.00
Scrap	14.00	14.00	90.00	95.00	3.00	2.00
Coal and coke	51.00	42.00	. . .	. . .	59.00	59.00
Other energy	22.00	20.00	45.00	44.00	10.00	11.00
Miscellaneous	88.00	90.00	68.00	63.00	94.00	94.00
Depreciation	10.00	11.00	10.00	9.00	16.00	17.00
Interest	4.00	8.00	8.00	12.00	18.00	17.00
Taxes	4.00	4.00	2.00	2.00	4.00	4.00
Total costs	381.00	362.00	282.00	267.00	303.00	295.00
Addendum						
Input prices						
Labor (dollars per man-hour)	19.50	22.50	16.85	17.50	11.50	11.70
Iron ore (dollars per ton)	45.00	40.00	. . .	. . .	27.00	24.25
Scrap (dollars per ton)	80.00	80.00	80.00	85.00	90.00	87.50
Coal (dollars per ton)	55.00	55.00	. . .	. . .	70.00	59.50
Electricity (dollars per kilowatt hour)	0.042	0.045	0.042	0.045	0.06	0.07
Yen per dollar	. . .	. . .	. . .	. . .	240	240
Efficiency measures						
Man-hours per ton	6.50	5.00	3.50	2.40	4.45	4.00
Yield to finished product (percent)	80	88	93	94	90	93

Sources: Table A-1; data provided to the authors by various producers; data from American Iron and Steel Institute; and Peter F. Marcus, Karlis M. Kirsis, and Donald F. Barnett, *World Steel Dynamics: The Steel Strategist #11* (New York: Paine Webber, 1985). Costs are based on 90 percent utilization of capacity. Tons in all tables are net tons unless otherwise noted.

operational improvements have resulted in more and more output with the same or smaller crews.

By contrast, no new plants have been built in the integrated sector since the early 1960s, and existing plants have been only partly rationalized to reflect prevailing economic conditions. Limited investment funds have been sprinkled over a large number of plants, many too small or poorly located for the changing steel market. And integrated firms have suffered from the restrictive union work rules built up over decades of collective bargaining.

Minimills have not been saddled with generations of accumulated management-labor antagonisms that have culminated in rigid work rules, low morale, and aggressive bargaining over wages. Integrated plants pay higher wages and get less from their workers in terms of productivity than the minimills.

Lower Wages

Most minimill companies have not been organized by the United Steelworkers of America. Since their plants are scattered around the country, often in small towns in the West and South, their wage rates reflect a variety of local labor-market conditions. Even the largest of the minimills, however, pay wages that are considerably lower than those at the major integrated companies. Total compensation in 1985 for the larger minimills was rarely more than $17.50 per hour, compared with $22.80 for the average integrated company.[2]

Some of the minimill companies, such as Nucor, have also been able to structure their compensation systems to provide substantial incentives for productivity. Thus, even the $17.50 hourly rate is achieved only with high productivity.

Product Specialization

Minimill companies differ in another important respect from their larger, integrated competitors. Each plant typically produces a very narrow range of products, and both the location and product mix of plants are designed to meet specific market requirements. Most of the companies produce a narrow range of carbon bars, reinforcing bars, and structural shapes from continuously cast billets or blooms. Some, such as Raritan River, produce only wire rods while others, such as Quanex and Newport, specialize in tubular products.

Reliance on a standardized, simple product line has been an important contributor to plant efficiency. Larger, integrated companies often produce a wide array of bar products or sheet products in a single plant, but their crude steel and primary rolling facilities are not easily adapted to vary their production of semifinished shapes as demand changes. The minimills, on the other hand, have been able to produce at maximum

2. American Iron and Steel Institute, *Annual Statistical Report, 1985* (Washington, D.C.: AISI, 1986), table 6; and authors' interviews with producers.

efficiency because their entire plant is geared to a very narrow range of output.

The risk in the minimill strategy is obvious; a fall in demand for a single product, such as small structural shapes, can be devastating. Indeed, firms like Florida Steel felt the full brunt of the slowdown in construction in 1982. On the other hand, the idling of minimill capacity is less expensive than the shutting down of huge integrated facilities.

Geographic Specialization

Transportation costs loom large for a relatively low-value product such as steel. As the market for steel moved west in the early twentieth century, steel makers moved to Chicago and even to Colorado and Utah to serve it. In lightly populated areas, the output of a 3 million- or 4 million-ton integrated plant must be shipped substantial distances to be marketed. Moreover, as markets decline and product demands shift, the large companies are forced to ship both raw materials and products hundreds of miles from one plant to another for processing. For instance, United States Steel ships coke from Pittsburgh to Cleveland and Philadelphia, slabs from Texas to Chicago, coils from Utah to California, and plates from Houston to Cleveland. Bethlehem Steel and National Steel Corporation also regularly ship such products between plants at huge transportation-cost penalties.

The earliest minimills that catered to specific regions, offering a fairly broad product mix of small bar shapes, merchant bars, and reinforcing bars, were protected from integrated mills' competition by high transport costs. The market mills that came later have a narrower product range—as little as two sizes of one product—and low production costs that allow them to ship to a wider geographic area. The superior efficiency of market mills like Raritan River and Chaparral gives them an advantage over other minimills and poses a serious challenge within the industry to integrated producers and importers.

Minimills clearly have an enormous locational advantage over their integrated rivals. Since they are not hostage to iron ore and coal supplies, they may go wherever electricity is available at a reasonable cost and there is a local market for 200,000 to 300,000 tons of a single product. Scrap is ubiquitous; hence, minimills are located in the New York area (Raritan River), near Chicago (Northwestern Steel and Wire), and in such remote locations as Plymouth, Utah, and Jackson, Mississippi.

The small producers can fulfill local demand for reinforcing bars, small structural shapes, other bar products, or wire rods without having to ship their products great distances while still obtaining scrap of reasonable quality in sufficient quantities. In addition, their small scale and simple plants do not result in the severe imbalances in facilities that arise in the integrated steel companies' coking, pig iron, steel-making, primary rolling, and strip mill facilities. Imbalances become a major problem for integrated producers if demand is too weak to allow them to utilize all facilities at their full capacity, as has been the case since the mid-1970s. Minimills not only have fewer facilities to balance but much smaller operating units to bear the cost penalty of falling production.

Low-Cost Scrap

Scrap is derived from yield losses in steel production and the fabrication of steel products and from the recycling of discarded steel-bearing articles such as automobiles, appliances, and even buildings. The recycled products are generally called obsolete scrap. Improved techniques for recycling discarded materials have greatly increased the return of obsolete scrap in the past few decades. Large crushers of automobiles and appliances have helped, and concern over the degradation of the environment has also contributed to this increased supply.

Scrap has always been relatively plentiful in the United States. Over time, the cost advantage of this ferrous material has in fact widened, compared with that of iron ore (except in 1973–74). In the 1970s, integrated producers gambled on developing high-cost domestic sources of iron ore to meet projected shortages, but over the past decade scrap prices have declined, and U.S. iron ore costs have risen well above world market levels. Low-cost scrap has been a major benefit to the expansion of scrap-based facilities in the United States. In other countries—say, Japan—where scrap is scarce and iron ore cheap, the advantages of minimill production are less pronounced.

In the final analysis, the success of minimills as a group (though many have failed) has involved a dynamic process of adjustment to new technology. The minimills began with strengths in the use of lower-cost labor and scrap but with scale disadvantages and a fairly inefficient, untried technology. Once they had established a bridgehead in low-grade steel, the next generation of mills incorporated operating improvements that more than offset any increase in capital costs, improving their competitive stance in their regional markets. As their competitiveness improved, the sophistication of their product range broadened, and their continued specialization allowed them to exploit the economies of scale. Each new generation of mills aimed at a new market, was more cost competitive, and used a superior technology. Eventually minimills' cost advantages became so great that they could aim for nationwide markets, launching a frontal assault on the integrated firms.

Production Costs

Unlike their integrated rivals, minimills have been successful in capturing selected markets by using successive minor changes in technology and operating practices that gradually lowered the real costs of steel making. The cost of producing wire rods for typical minimill producers, estimated in table 2-1, are much lower than those of their integrated rivals in the United States and even below the costs for integrated Japanese producers.

Between 1960 and 1980, the gap in costs between integrated U.S. producers and minimills widened as integrated costs rose. Though costs in both sectors fell between 1981 and 1985, the gap remained. Japanese integrated costs, in U.S. dollars, fell in those years largely because of lower iron ore and coal prices (due to the strong dollar) and some improvements in efficiency.

The low labor costs of U.S. minimills are especially noteworthy— about the same as in Japanese integrated plants despite substantially higher hourly labor costs in the United States. Similarly, because they use scrap rather than iron ore, U.S. minimills' energy costs are much lower than those of their integrated competitors. It is only in the cost of metallics (scrap versus iron ore) that minimills suffer a disadvantage relative to integrated firms.

The data on capital charges are deceptive. Minimill capital costs per ton are far below those of new integrated plants. The differences in finance charges shown in table 2-1 reflect differences in the ages of facilities and in accounting conventions and in debt-to-equity ratios. Assuming the same accounting procedures and debt-to-equity ratios were used, however, financial charges per ton of product in new minimills would be much less than those in new integrated mills. New integrated plants would have lower operating costs, but their total costs would be

Table 2-2. *Cost of Producing Wire Rod in an Efficient Minimill, Five Countries, 1985*

Item	United States	West Germany	Japan	South Korea	Brazil
	Dollars per ton of finished product				
Operating costs	224.00	216.00	232.00	235.00	212.00
Labor	34.00	23.00	25.00	12.00	14.00
Scrap	93.00	92.00	95.00	102.00	81.00
Iron	. . .	. . .	. . .	. . .	28.00
Energy	41.00	35.00	51.00	54.00	23.00
Miscellaneous	56.00	66.00	61.00	67.00	66.00
Depreciation	12.00	11.00	13.00	12.00	12.00
Interest	18.00	18.00	11.00	10.00	11.00
Taxes	2.00	2.00	2.00	1.00	1.00
Total costs	256.00	247.00	258.00	258.00	236.00
Addendum					
Input prices					
Labor (dollars per man-hour)	17.50	11.50	11.70	2.40	2.50
Scrap (dollars per ton)	85.00	85.00	87.50	94.00	90.00
Electricity (dollars per kilowatt hour)	0.045	0.040	0.063	0.060	0.017
Exchange rate (national unit per dollar)	. . .	2.90	240	800	8,500
Efficiency measures					
Man-hours per ton	1.95	2.00	2.10	2.60	5.50
Electricity use (kilowatt hours per ton)[a]	485	465	460	500	470
Yield to finished product (percent)	96	96	96	95	92

Sources: Tables 2-1, A-2. Costs are based on 90 percent utilization of capacity.
a. Electric furnace and caster only.

even higher than the 1985 integrated costs shown in table 2-1 (see chapter 3).

International Competitiveness of Minimills

In spite of the large difference in their hourly costs for labor, U.S. minimills were competitive with Japanese integrated producers of wire rods even at 1985 exchange rates. But how do U.S. minimills compare with minimills in other countries? Overall production costs for efficient minimills in Brazil, South Korea, Japan, West Germany, and the United States are remarkably similar (see table 2-2). Though U.S. labor costs are higher (despite higher productivity) than other countries', this is offset by lower metallic (scrap) costs. Thus the U.S. disadvantage in

Table 2-3. *Cost of Producing Wire Rod in a Representative Minimill and a Low-Cost Minimill in the United States, 1985*

Item	Representative mill	Low-cost mill
	Dollars per ton of finished product	
Operating costs	244.00	224.00
Labor	42.00	34.00
Scrap	95.00	93.00
Energy	44.00	41.00
Miscellaneous	63.00	56.00
Depreciation	9.00	12.00
Interest	12.00	18.00
Taxes	2.00	2.00
Total costs	267.00	256.00
Addendum		
Input prices	a	a
Efficiency measures		
Man-hours per ton	2.4	1.95
Electricity use (kilowatt hours per ton)[b]	470	425
Electrode use (pounds per ton)	9	7
Tap-to-tap time (hours)	1.5	1.25
Yield to finished product (percent)	94	96
Wire rod rolling speed (feet per minute)	15,000	20,000

Sources: Tables A-1, A-2; data provided by producers. Costs are based on 90 percent utilization of capacity.

a. See under United States in table 2-2.

b. Electric furnace and caster only.

labor is not as damaging for minimills as for integrated producers. And capital charges are small enough and similar enough (most minimills are of similar age because they are short-lived) not to cause major differences in costs—which would not be true for integrated producers.

The Trend in Minimill Costs

The secret of minimill success lies in both input costs and technical efficiency. Minimills pay less than integrated firms for labor, metallic inputs, and energy, but they are also more efficient in the use of these inputs. Moreover, this efficiency has been improving, as the cost data on production of wire rods demonstrate (see table 2-3).[3] Minimills' use

3. Wire rod is used for comparative purposes because it has been produced by both minimills and integrated firms; minimills' costs for producing reinforcing bars and special quality bars and the prospective costs of producing seamless tubes are presented in appendix table A-7.

of electricity, electrodes, and refractories has been reduced because of dramatic improvements in electric furnace technology. Ultrahigh power, water-cooled panels, and oxy-fuel burners conserve energy and materials; furnace tap-to-tap times in 1985 were less than 1.5 hours for the representative mill, compared with about 5.0 hours a decade earlier. Raw steel costs in minimill operations have been dramatically lowered.

And the yields from raw to finished steel are higher in minimills than in integrated plants because of continuous casting and the narrow range of goods produced. Minimills have also installed newer high-speed rolling mills—in 1985 typical wire rod mills were operating at 20,000 feet per minute, compared with 15,000 feet per minute three years earlier. Such improvements reduce energy use, yield loss, and capital costs per ton of input.

Perhaps the single greatest improvement in minimill performance has been in labor use. In 1985 a typical minimill producing wire rod used about 2.4 man-hours per ton, the typical integrated mill about 5.0 man-hours (see table 2-1). Lower labor use makes a dramatic difference in the total costs of production, as a comparison of 1981 with 1985 makes evident. In just four years, average man-hours per ton fell from 3.5 to 2.4 (reflecting the building of new minimills and improvements in older plants).

The data in table 2-2 demonstrate that an efficient U.S. minimill and an efficient Japanese minimill have similar costs at an exchange rate of 240 yen to the dollar. In 1981 U.S. and Japanese minimills were equally efficient, but by 1985 larger, newer mills in the United States, such as Raritan River, had given efficient U.S. mills a temporary lead in productivity (1.95 versus 2.10 man-hours per ton in 1985). However, progress in this sector is so rapid that the U.S. lead could disappear in a very short time.

The low-cost mill portrayed in table 2-3 is still not considered state of the art in terms of efficiency and production costs. More modern minimills are expected to realize still greater cost savings, sufficient to justify the capital expenditures necessary to modernize them. Clearly, minimills are constantly running to catch a state-of-the-art challenge that is forever two paces ahead. This is in stark contrast to integrated production in the United States, which is locked into older plants that cannot be revamped easily to reflect the state of the art in either blast furnace or basic oxygen furnace technology.

Exchange Rates

The international cost comparisons in table 2-2 are based on exchange rates in mid-1985. But as the value of the U.S. dollar against the currencies of the United States' more important trading partners changes, the relationship of U.S. costs to those of foreign minimills will vary. For example, if the value of the dollar is equal to 170 yen, efficient U.S. minimills will have a $69 per ton cost advantage over efficient Japanese producers:

Type of cost	U.S. costs (dollars)	Japanese costs (dollars) at a yen-dollar rate of			
		150	170	210	240
Operating	224	312	289	251	232
Financial	32	41	36	30	26
Total	256	353	325	281	258

Similarly, a dollar worth 2.20 deutsche marks will give U.S. producers a $49 per ton cost advantage over efficient German plants:

Type of cost	U.S. costs (dollars)	West German costs (dollars) at a DM-dollar rate of			
		2.00	2.20	2.60	2.90
Operating	224	280	263	236	216
Financial	32	46	42	36	31
Total	256	326	305	272	247

At this low value of the dollar, U.S. producers face no competition in the U.S. market for wire rods from most of the world's other minimills.

The Differential in Liquid Steel Costs

If minimills are so much more efficient than integrated firms in producing small-diameter goods, such as wire rods, why did they not take over the market for these products decades ago? Part of the explanation lies in the recency of technological advances in melting scrap and forming steel in continuous casters. But the pace of minimill growth has been greatly assisted by the decline in scrap prices since 1974 and the strategic errors that the larger firms made, particularly in their investments in iron ore during the 1970s.

Scrap

Minimills had a very small share of the U.S. industry in the early 1970s when scrap prices were high relative to the integrated firms' cost of producing liquid steel from coal, iron ore, and limestone. Integrated firms can combine the hot metal from their blast furnaces with scrap in basic oxygen furnaces whose operating costs are typically lower than the costs that minimills face in melting scrap in an electric furnace. Obviously, the relative prices of scrap and hot metal are a crucial determinant of the position of minimills.

Integrated firms can use scrap for as much as a third of the metallics they charge into a basic oxygen furnace without preheating or otherwise changing their processing; the remaining two-thirds is generally hot metal. If scrap prices are far below the costs of producing hot metal, minimills will have a decided advantage. The degree of advantage can be determined by letting

(1) $$M_m = C_m + 1.1 \, PSCRAP$$

and

(2) $$M_i = C_i + 0.80 \, HMCOST + 0.34 \, PSCRAP,$$

where M_m represents the cost of making liquid steel at minimills and M_i the cost at integrated plants using basic oxygen furnaces, C_m the cost of operating an electric furnace, C_i the cost of transforming hot metal and scrap into liquid steel in a basic oxygen furnace, $PSCRAP$ the price of scrap, and $HMCOST$ the cost of producing molten pig iron from coal, iron ore, and limestone. If integrated and minimill steel-making costs are to be equal, the price of scrap can be obtained by setting equation 1 equal to equation 2, which can be reduced to:

(3) $$PSCRAP = 1.05 \, HMCOST - 1.32 \, (C_m - C_i).$$

At 1985 costs, a basic oxygen furnace is approximately $27 per ton cheaper to operate than an electric furnace, and it costs about $174 (including capital costs) to produce a ton of hot metal. If minimills are to have costs equal to integrated firms' costs, the price of scrap should be about $147 per ton, or about 84 percent of hot metal costs.

In 1973, scrap sold for an average of about 77 percent of hot metal costs. Since then, its price has fallen dramatically relative to hot metal costs, reaching a low of $56.37 per ton in 1982, or just 31 percent of hot

Table 2-4. *Cost of Hot Metal and Scrap, 1973–85*
Dollars per ton

Year	Hot metal	Scrap[a]
1973	67.40	51.66
1974	84.01	96.87
1975	102.06	64.15
1976	109.95	69.41
1977	110.39	56.38
1978	128.46	68.16
1979	141.88	87.42
1980	158.30	81.35
1981	174.63	81.70
1982	180.98	56.37
1983	176.90	65.59
1984	177.35	76.84
1985	174.00[b]	66.27[c]

Sources: Marcus and others, *World Steel Dynamics*, table 19; David J. Joseph Co., "The Relationships between Scrap Prices, Steel Production, Purchased Scrap Receipts, and Scrap Exports 1954–1983" (Cincinnati: The Company, 1984).

a. Price of number 1 heavy melting scrap.
b. Based on projections for 1985–87.
c. Estimate.

metal costs (see table 2-4). Even with the economic recovery of 1983–85, scrap prices were still less than 40 percent of hot metal costs in 1985.

The reason for the sharp imbalance in scrap prices and hot metal costs is quite simple: world steel consumption has been essentially flat since 1974. When steel output is not growing, the potential supply of scrap will grow relative to the metallics required in steel production. Technological improvements in recovery methods and increased supplies of obsolete scrap assure this result. Shredders and other machines for fragmenting and sorting automobile hulks have added to the scrap supply. Moreover, the supply in 1984 was much greater than in 1974 because obsolete scrap comes from goods produced as much as ten or twenty years earlier (see chapter 5). Thus the slowdown in growth reduced steel production far more than it reduced the supply of recycled scrap.[4]

Iron Ore

In the 1970s, the cost of iron ore in the United States rose rapidly; U.S. firms were paying as much as 58 percent more for ore than their

4. Marian Radetzki and Carl Van Duyne, "The Demand for Scrap and Primary Metal Ores after a Decline in Secular Growth," Research Paper 58 (Williams College, Department of Economics, June 1983).

Table 2-5. *Delivered Cost of Iron Ore in Japan and the United States, Various Years, 1960–85*
Dollars per metric ton

Year	United States	Japan
1960	12.29	14.20
1965	13.01	13.42
1970	14.39	11.84
1975	26.44	16.70
1980	34.50[a]	27.82[b]
1985	46.00[a]	27.00[b]

Sources: U.S. Federal Trade Commission, *The United States Steel Industry and Its International Rivals: Trends and Factors Determining International Competitiveness* (FTC, 1977), p. 117; Donald F. Barnett and Louis Schorsch, *Steel: Upheaval in a Basic Industry* (Ballinger, 1983), p. 303; Marcus and others, *World Steel Dynamics*, table 26.
a. Cost of pellets delivered to Lake Erie from U.S. mines.
b. Cost of pellets delivered to Yokohama from Australia.

Japanese rivals by 1975 (table 2-5). Nevertheless, the U.S. integrated firms continued to expand their iron ore mining and their facilities for making pellets through the 1970s (see chapter 3). By 1980, ore could be imported from Australia or Brazil into the Great Lakes at a delivered price below the cost of U.S. ore.[5] The price premium on U.S. iron ore developed at a time of decline in the real price of scrap, saddling the integrated firms with a large and growing cost disadvantage in raw materials. This disadvantage was compounded by the firms' decisions in the 1960s and 1970s to build large basic oxygen furnaces and hot-strip mills and their failure to diagnose a secular downward shift in the demand for steel.

The Effect of Minimills on Prices

The entry of new minimills in the 1970s and the relentless downward pressure on costs in their segment of the industry had predictable effects on prices. Products that have been dominated by the minimill producers have had far more modest price increases than those dominated by integrated producers, as the weighted, inflation-adjusted indexes of prices indicate (figure 2-1). Both indexes exhibit a sharp rise in real prices in 1974, but the index for minimill products fell much more rapidly from its 1974 peak than that of integrated firms' output. By 1984, the real

5. Donald F. Barnett and Louis Schorsch, *Steel: Upheaval in a Basic Industry* (Ballinger, 1983), pp. 301–03.

Figure 2-1. *Real Prices for Steel Produced by Integrated Firms and Minimills, 1971–84*

Dollars per net ton (1972 dollars)

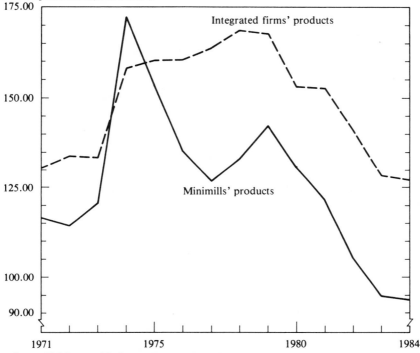

Source: U.S. Bureau of the Census, *Current Industrial Reports: Steel Mill Products, 1984*, MA-33B (Census Bureau, 1985), and earlier issues.

price of minimill products was 24.2 percent below its 1971 level, but integrated products were only 2.6 percent lower than their 1971 level. The entry of Georgetown Steel, Chaparral, Raritan River, and Bayou in the 1970s undoubtedly contributed to the substantial decline in minimills' real prices.

The effects of technological progress and increased market competition on prices can be ascertained more precisely by an empirical investigation of the determinants of individual steel prices. In a competitive or noncompetitive market, prices should depend on how costs and demand vary. Specifically, the variable costs (V) of the ith product are a function of an index of the price of materials ($MATCOST$), the wage rate ($WAGE$), and a time trend ($TIME$) in the year t, which captures productivity growth:

(4) $$V_{it} = f(MATCOST_t, WAGE_t, TIME_t).$$

The markup of prices (P) over costs should be a function of the pressure of demand on capacity. This may be measured by the domestic industry's utilization of production capacity ($CAPUTIL$). In addition, prices are affected by periodic bursts of trade protection, which can only be captured by dummy variables (D_{jt}) for the jth year of trade protection. Let two dummy variables represent the effect of binding voluntary restraint agreements in 1971–72 and two others represent the effect of binding trigger prices in 1978–79. The effect of minimill competition can then be estimated by using annual time series data for 1965–82 to estimate the following equation:

(5) $$\text{Log } P_{it} = a_0 + a_1 \log MATCOST_t + a_2 \log WAGE_t + a_3 TIME_t$$

$$+ a_4 \log CAPUTIL_t + \sum_{j=1}^{J} a_{j+4} D_{jt} + u_t.$$

The effect of minimills should be reflected in equation 5 in a relatively high rate of productivity growth; therefore, the absolute value of the estimate of a_3 that fits best should be higher for minimill products than for integrated products. And for products where minimills have recently entered the market—such as wire rods—there should be a break in prices, which should be reflected by the appearance of negative residuals after 1975. Following are the estimates of productivity growth in 1965–82 and the difference between actual and predicted prices in 1976–82 for four products dominated by integrated firms, one that has recently shifted to the minimills (wire rods), and one long-established minimill product (reinforcing bars):

Item	Hot-rolled sheet	Cold-rolled sheet	Plates	Structural shapes	Wire rod	Reinforcing bars
Sum of residuals from equation 5, 1976–82	−1.1	+1.7	−3.1	−0.7	−9.1	−9.4
Estimated productivity growth a_3, 1965–82 (percent)	3.9	2.0	0.9	−0.1	2.9	10.0

The results graphically confirm our expectations about productivity growth. Minimill products evidence annual rates of productivity growth of 2.9 percent and 10.0 percent, while two of the integrated products show no significant productivity growth and the other two evidence annual productivity growth of 2.0 and 3.9 percent. Wire rods exhibit large negative residuals after 1975 (when Raritan River and Georgetown's Texas plant began production). And reinforcing bars also show a large price break after 1975. Minimills appear to have driven prices down through competitive pressures arising in part from their greater productivity growth. A similar pattern is likely to emerge in the other products when the minimill assault is extended to them. The four integrated products, by contrast, show very little break in prices after 1975.

Summary

The development of new, lower-cost production technologies for electric furnaces and continuous billet casters has stimulated the startling recent growth in the minimill sector of the U.S. steel industry. The new plants' costs are decidedly lower than those of their integrated rivals.

International competition has proved devastating to the integrated companies, whose costs were rising relative to their more efficient foreign rivals' through 1985. The production costs of U.S. minimills have been very close to those in foreign minimills, and they have been declining in recent years. The minimills have been able to generate much larger productivity gains than the integrated firms and they have enjoyed much lower raw materials costs. They have been able to reduce prices and meet foreign competition—albeit at compressed profit margins in the 1980s—while the integrated firms often have not. As the U.S. dollar has fallen in 1986, the minimills have become virtually invulnerable to foreign competition, and the integrated producers have become more cost competitive.

III

The Decline of the Integrated Sector

For decades the U.S. steel industry's difficulties have been charged to the big producers' failure to invest in modern, large-scale technology. In the 1960s, the industry was criticized for failing to adopt the basic oxygen furnace as rapidly as the Europeans or the Japanese.[1] In the 1970s, it was criticized for failing to adopt continuous casting and for continuing to use outmoded blast furnaces.[2] But would the industry have been better able to compete had it moved swiftly to modernize?

Unexpected Pressures in the 1960s and 1970s

Perhaps the most traumatic event in the history of the U.S. steel industry since World War II was the strike of 1959. For nearly four months the industry was shut down, and as a result imports became a significant factor for the first time. During the next decade, imports were to rise erratically to 18 million tons after having averaged less than 1.5 million tons in the three years preceding the strike (table 3-1). For the first time, the U.S. steel industry faced important foreign competition.

The early 1960s were a period of very slow growth in steel output in the United States, partly because of rising imports. Output did not rise above its 1956 peak until 1964. Later, with the economic expansion of the Vietnam War period, demand for steel grew rapidly and the U.S. industry moved to replace its open hearth furnaces and outmoded rolling mills. By 1970, nearly 50 percent of the industry's raw steel output was from large, new basic oxygen furnaces (BOFs), and six new large-scale hot-strip mills for finishing the various steel shapes had been built.[3] By

1. Walter Adams and Joel B. Dirlam, "Steel Imports and Vertical Oligopoly Power," *American Economic Review*, vol. 54 (September 1964), pp. 626–55.
2. For assessments of the industry's position in the 1970s, see Robert W. Crandall, *The U.S. Steel Industry in Recurrent Crisis* (Brookings, 1981); and Donald F. Barnett and Louis Schorsch, *Steel: Upheaval in a Basic Industry* (Ballinger, 1983).
3. American Iron and Steel Institute, *Annual Statistical Report, 1984* (Washington, D.C.: AISI, 1985), table 27.

Table 3-1. *Production, Shipments, and Imports of Raw Steel in the United States, 1956–85*

Year	Raw steel production	Total net shipments	Imports
	Millions of tons		
1956	115.2	83.3	1.3
1957	112.7	79.9	1.2
1958	85.3	59.9	1.7
1959	93.4	69.4	4.4
1960	99.3	71.1	3.4
1961	98.0	66.1	3.2
1962	98.3	70.6	4.1
1963	109.3	75.6	5.5
1964	127.1	84.9	6.4
1965	131.5	92.7	10.4
1966	134.1	90.0	10.8
1967	127.2	83.9	11.5
1968	131.5	91.9	18.0
1969	141.3	93.9	14.0
1970	131.5	90.8	13.4
1971	120.4	87.0	18.3
1972	133.2	91.8	17.7
1973	150.8	111.4	15.2
1974	145.7	109.5	16.0
1975	116.6	80.0	12.0
1976	128.0	89.4	14.3
1977	125.3	91.1	19.3
1978	137.0	97.9	21.1
1979	136.3	100.3	17.5
1980	111.8	83.9	15.5
1981	120.8	88.5	19.9
1982	74.6	61.6	16.7
1983	84.6	67.6	17.1
1984	92.5	73.7	26.2
1985	88.3	73.0	24.3
	Annual rate of growth (percent)		
1956–66	1.5	0.8	23.6
1966–73	1.7	3.1	4.9
1973–79	−1.7	−1.8	2.5
1973–81	−2.8	−2.9	3.5
1979–85	−7.2	−5.3	5.5

Source: American Iron and Steel Institute, *Annual Statistical Report, 1985* (Washington, D.C.: AISI, 1986), tables 1A, 1B, and earlier issues. Tons in all tables are net tons unless otherwise noted.

1982 open hearth steel making had fallen to less than 10 percent of industry output, and only United States Steel had kept a significant amount of open hearth capacity.

The conversion to new furnaces and rolling mills occurred during a period of considerable optimism in the U.S. industry. Between 1964 and 1974, raw steel production was rarely less than 130 million tons per year, and the strong rise in demand in the early 1970s led the steel industry to expect even more rapid growth.[4] Unfortunately, the 1973–74 level was a peak. Raw steel output has remained below 100 million tons per year since 1981.

The rapid market growth of the late 1960s and early 1970s—reinforced by the steel shortages of 1973–74—fueled optimism and led the integrated firms to believe that they should maintain or expand the capacity of their existing plants. Large basic oxygen furnaces were installed to replace the open hearths, even in plants whose finishing facilities did not require large-scale output. For instance, Bethlehem Steel put three vessels with a capacity of 310 tons each in its Lackawanna plant and later attempted to compensate for the imbalance in finishing facilities by building a large bar mill there. Unfortunately, the new furnaces were not equipped with modern billet casters to feed the bar mill, and the steel-making facilities were eventually closed.[5]

An even more graphic example is Kaiser's late 1970s decision to build a basic oxygen facility capable of producing 2.8 million tons of molten steel a year and a continuous caster that could produce 700,000 tons of slabs annually. These facilities, which cost $240 million, went into operation in 1978 and were closed in 1983. The entire plant, including rolling mills, was subsequently sold for $110 million. Only the rolling mills have been used by the new owners, to roll slabs imported from Brazil and Europe.

By the mid-1970s, the installation of the large new furnaces and rolling mills had largely been completed, but most of the integrated plants still depended on ingot casting and primary rolling rather than continuous casting. Moreover, their coke ovens and blast furnaces could not produce pig iron efficiently for the basic oxygen furnaces, which generally use a larger ratio of hot metal to scrap than the open hearths they replaced.

4. AISI, "Financing Capital Expenditures: A Critical Concern of the American Steel Industry," statement prepared for the Council on Wage and Price Stability, October 30, 1975, pp. 3–7.

5. Apparently, the cost of producing raw steel was too high to justify shipping semifinished steel to other plants on the Great Lakes.

The Japanese meantime had pioneered large blast furnaces and by 1975 had a large share of the world's most efficient furnaces. The U.S. blast furnaces, by contrast, were much smaller and less efficient. Also, by 1975, 31 percent of Japanese steel was continuously cast while only 9 percent of U.S. steel was.[6] Modernity and efficiency were more than simply a matter of toting up BOF capacity.

The U.S. producers' decision to install huge new furnaces and hot-strip mills at most of their plants has proved to be an unfortunate legacy. They could only justify that investment by investing enormous sums in blast furnaces, continuous casters, and more modern rolling mills. Indeed, the centerpiece of the integrated firms' strategy in the 1970s was the attempt to "round out" capacity.[7] Their strategy might have worked if demand had continued to grow and inflation had persisted. Strong growth and a very weak dollar might have allowed the industry to modernize its old plants, and American producers might have remained reasonably competitive with the new exporters in developing countries, the Japanese, and the better European producers. Moreover, the price of scrap might have stayed high enough to slow the expansion of minimills.

Obviously, events were unkind to the integrated companies. Demand sagged badly and the dollar appreciated after a brief decline in 1978–79. With weak demand, the integrated companies found it increasingly difficult to maintain their investment plans. As a result, most plants suffered from imbalances and inefficiencies in coking, smelting (blast furnaces), casting, or rolling. The integrated producers could not operate their new large-scale furnaces efficiently, and competition often forced them to ship their reduced output substantial distances.

The effect of the typical integrated firm's decision to build new basic oxygen furnaces was to increase its efficient scale of production. But the additional capital costs of the new facility would raise the unit cost of production substantially above the unit cost in the old open hearths if capacity was less than fully utilized. The crossover point for lowering unit production costs in switching from *old facilities* to *new facilities* is depicted in figure 3-1. For plants that are unable to replace older blast furnaces and improve their other facilities, the unit cost of production with a new basic oxygen furnace could be as high as that shown for

6. International Iron and Steel Institute, *Steel Statistical Yearbook, 1984* (Brussels: IISI, 1984), p. 10.

7. See, for example, the discussion in AISI, *Steel at the Crossroads: The American Steel Industry in the 1980's* (Washington, D.C.: AISI, 1980), chap. 5.

Figure 3-1. *Average Production Costs in an Integrated Plant and in a Minimill*

Costs (dollars per ton)

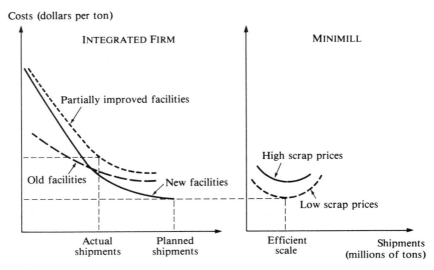

partially improved facilities—even higher, that is, than that of the old open hearth facility.

In the 1960s and 1970s, the minimills were becoming competitive with integrated firms. Their costs for many products might have remained above those at integrated plants had the economic climate been different. In fact, the slow growth in demand and the high value of the dollar depressed scrap prices substantially, thereby lowering the minimill cost for *every* product. Slow growth thus increased the minimill cost advantage substantially, as can be seen in the downward shift in the minimill unit cost in figure 3-1. Even if minimills' unit costs at *high scrap prices* are relatively similar to the unit costs of integrated plants with *new facilities* operating at minimum efficient scale, slow economic growth and imbalances at the integrated plants have driven a substantial wedge between costs in the two sectors in the 1980s.

Escalating Wage Rates

Another unfortunate consequence of the overoptimism of the 1970s was the industry's willingness to accede to large demands from the steelworkers in exchange for labor peace. After the long strike in 1959, wage increases in the steel industry were moderate. With the Vietnam

Figure 3-2. *Total Compensation for Production Workers in Steel and All Manufacturing Industries, 1967–84*

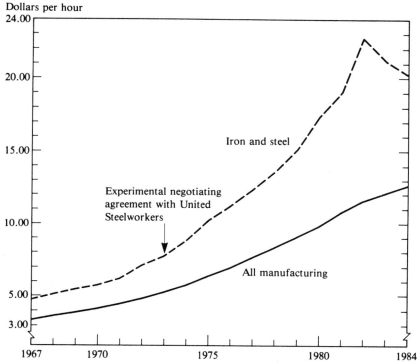

Dollars per hour

Source: U.S. Department of Labor, Bureau of Labor Statistics, unpublished estimates of total compensation, January 1985.

War boom, however, the industry found itself faced with escalating demands from the union, to which it acceded in 1968 and 1971. In 1973, when total compensation for iron and steel production workers was about 50 percent above the average for all manufacturing industries (figure 3-2), the industry reached an "experimental negotiating agreement" with the United Steelworkers of America that guaranteed the workers at least 3 percent increases in real wages each year in return for an agreement not to strike. By 1982, total compensation in iron and steel had risen to approximately double the average for all manufacturing.

These increases in compensation were not offset by increases in labor productivity. As a result, unit labor costs in the steel industry rose by 220.4 percent between 1972 and 1982, while those in all manufacturing rose by 102.2 percent (table 3-2). In short, a sharply declining industry

Table 3-2. *Index of Unit Labor Costs in Steel and All Manufacturing Industries, 1964 and 1972–84*
Base year 1977 = 100

Year	Steel	All manufacturing
1964	44.3	58.1
1972	60.2	73.0
1973	60.2	74.2
1974	68.7	84.1
1975	91.2	91.5
1976	93.8	94.6
1977	100.0	100.0
1978	103.1	107.3
1979	115.6	117.0
1980	134.4	130.5
1981	137.6	138.4
1982	192.9	147.6
1983	152.5	146.4
1984	131.8	146.5

Source: Bureau of Labor Statistics, unpublished data. Costs are based on all employees.

was faced with increases in labor costs that were double the manufacturing average over this ten-year period.

Why wage rates should have risen so rapidly in a declining industry is far from clear. One suggestion is that the union may have been playing an endgame, extracting as much of the quasi-rents as possible from the industry since reinvestment in productive capacity had become unprofitable.[8] On the other hand, the important milestones in the industry's labor negotiations were 1973 and 1974, when the industry's future appeared better than it had at any time since World War II.

In 1983, the industry was able to gain some concessions from the United Steelworkers. Total compensation fell from $22.72 in 1982 to $21.14 in 1983 and to $20.28 in 1984, in part because of the lower wages of the minimill companies and wage concessions granted to the integrated producers.[9] Unit labor costs fell substantially in 1983 and 1984 as the troubled integrated companies closed facilities, reduced work crews, and shed salaried personnel. Nevertheless, total compensation for

8. Colin Lawrence and Robert Z. Lawrence, "Manufacturing Wage Dispersion: An End Game Interpretation," *Brookings Papers on Economic Activity, 1985:1*, pp. 47–106.

9. Bureau of Labor Statistics, unpublished estimates, January 1986.

member firms of the American Iron and Steel Institute remained 66 percent above the manufacturing average. In the equally troubled European steel industries, total compensation is generally less than 30 percent above the manufacturing average.[10]

Investment Errors in the 1970s

The integrated firms were very slow to recognize that circumstances were changing after the world steel shortage of 1973–74. In 1975 and 1976, the American Iron and Steel Institute forecast a rise in demand to 132 million tons of finished steel by 1983, requiring a U.S. capacity of 185 million tons of raw steel, or 30 million more than the existing capacity.[11] Even as late as 1980, the industry was forecasting that steel consumption in 1988 would be 134 million tons and suggesting that domestic firms should strive for 168 million tons of raw steel capacity.[12] Unfortunately, as the business cycle approached its peak in 1985, U.S. consumption was only 96 million tons and production only 88 million tons. Yet U.S. firms were still maintaining their nominal capacity to produce more than 130 million tons.

Had integrated steel makers known that there would be a decline in steel sales over the 1974–84 decade, they might have behaved quite differently. They might not have built many of their large new blast furnaces or basic oxygen furnaces. Nor would they have invested huge sums in iron ore mines and iron pelletizing facilities.

By 1970, the Japanese already had lower delivered iron ore costs than the American firms despite the fact that Japan's ore had to be imported from Australia and Brazil (see table 2-5). Despite the growing cost disadvantage of Great Lakes ore production, each of the integrated U.S. firms chose to make major investments in domestic iron ore projects in the 1970s.[13]

10. Ibid.
11. AISI, "Financing Capital Expenditures," table A-2.
12. AISI, *Steel at the Crossroads*, p. 2.
13. Bethlehem Steel built a 9-million-ton taconite pellet plant in Hibbing, Minnesota, in partnership with others. Republic, one of the participants in that project, obligated its plants to take 1.3 million tons per year. Jones and Laughlin expanded its mine and pellet plant at Tilden, Michigan, by a projected 4 million tons, through a partnership. It was obligated to take 1.44 million gross tons per year from the plant, and Wheeling-Pittsburgh Steel, another participant, was obligated to take 0.8 million tons per year. United States Steel expanded its mine and pellet plant at Mountain Iron, Minnesota,

Table 3-3. *Gross Investment in the Steel and Iron Ore Industries, 1965–82*

Year	Investment (billions of 1972 dollars)		Investment in iron ore as a percent of investment in steel
	Steel industry	Iron ore and pellet plants	
1965	1.95	0.40	20.51
1966	2.19	0.24	10.96
1967	2.38	0.22	9.24
1968	2.34	0.78	33.33
1969	1.95	0.70	35.90
1970	1.58	0.23	14.56
1971	1.15	0.41	35.65
1972	1.06	0.14	13.21
1973	1.22	0.21	17.21
1974	1.63	0.25	15.34
1975	1.71	0.29	16.96
1976	1.66	0.32	19.28
1977	1.59	0.41	25.79
1978	1.50	0.46	30.67
1979	1.45	0.48	33.10
1980	1.51	0.42	27.81
1981	1.51	n.a.	. . .
1982	1.14	n.a.	. . .
	Annual rate of growth (percent)		
1970–80	− .45	6.02	6.24

Source: U.S. Department of Commerce, Office of Business Analysis, unpublished data.
n.a. Not available.

Investment in iron ore and pellet plants, which had generally fallen to less than 20 percent of steel investment in the United States in the early 1970s, rose dramatically in the late 1970s (table 3-3). By 1978–79, it was nearly a third of steel investment despite the fact that world prices for iron ore were falling. By the early 1980s, iron ore from Brazil or Australia could compete in the Chicago area.

The integrated producers were investing in higher-cost domestic ore because they believed that future steel shortages would drive up Australian and Brazilian prices and that the guaranteed access to domestic ore

---

by 6.7 million tons per year, and National Steel expanded its iron ore pellet capacity from 6.5 million tons per year in 1975 to nearly 10.0 million in 1978. Inland Steel expanded its Minnesota pellet capacity by 2.9 million tons and its share of Empire Mine in Palmer, Michigan, by 1.3 million tons. (From annual reports of the companies.)

was insurance well worth the price. With the decline in world steel consumption, which set in precisely as these vast North American ore investments were being made, the demand for iron ore has plummeted. Most of the U.S. integrated companies now have vast iron ore facilities in Minnesota and Canada that will never be worth more than a fraction of their original cost.

A Struggle for Survival in the 1980s

As the 1980s unfolded, it was quite clear that the integrated companies had made a serious mistake based on their erroneous forecasts of steel demand. They had the capacity, mostly in large basic oxygen furnaces, to produce about 140 million tons of raw steel annually. They had kept far too many plants open in anticipation of demand growth that never occurred. With excessive investments in iron ore, coal, and raw steel capacity, they had no choice but to begin closing mills and selling off redundant raw material assets. Unfortunately, the iron ore facilities are worth very little, and even coal prices have lagged because of the strong dollar and the fall in real oil prices.

Cost Competitiveness

Integrated producers in the United States had substantially higher production costs in their more efficient plants in 1985 than the West German, Japanese, South Korean, or Brazilian producers (table 3-4). Operating costs in the more efficient flat-rolled steel plants of Japan, South Korea, and Brazil were approximately 30 percent lower than in efficient U.S. plants in 1985 before the decline of the dollar. The differences are due largely to labor costs, but they also reflect sharply higher metallic costs in the United States. At mid-1985 exchange rates, even a 25 percent reduction in the compensation of unionized steelworkers in the United States would only reduce the South Korean–U.S. labor cost differential from $104 per ton to $78. The higher operating costs in the United States are partially offset by higher capital costs in Japan, Brazil, and South Korea, which have newer plants than the United States. This provides only a minor offset and little solace. New plants can be built for less today in Japan and South Korea than in the United States.

Table 3-4. *Cost of Producing Cold-Rolled Coil in an Efficient Integrated Steel Firm, Five Countries, 1985*

Item	United States	West Germany	Japan	South Korea	Brazil
	Dollars per ton of finished product				
Operating costs	403.00	324.00	286.00	270.00	274.00
Labor	129.00	70.00	63.00	25.00	26.00
Iron ore	67.00	47.00	44.00	48.00	24.00
Scrap	18.00	11.00	. . .	. . .	. . .
Coal or coke	50.00	48.00	52.00	55.00	68.00
Other energy	24.00	22.00	15.00	24.00	27.00
Miscellaneous	115.00	126.00	112.00	118.00	129.00
Depreciation	24.00	24.00	29.00	77.00	27.00
Interest	12.00	15.00	27.00	14.00	80.00
Taxes	7.00	1.00	5.00	1.00	3.00
Total costs	446.00	364.00	347.00	362.00	384.00
Addendum					
Input prices					
Labor (dollars per man-hour)	22.50	11.90	11.70	2.85	2.90
Iron ore (dollars per ton)	40.00	26.00	24.25	25.00	12.50
Coal (dollars per ton)	55.00	58.00	59.50	59.00	60.00
Exchange rate (national unit per dollar)	. . .	2.90	240	800	8,500
Efficiency measures					
Man-hours per ton	5.75	5.85	5.35	8.20	9.00
Yield to finished product (percent)	78	80	89	82	80
Iron ore per ton of finished product	1.67	1.81	1.81	1.92	1.92

Sources: Table A-4; data provided to the authors by various producers. Costs are based on 90 percent utilization of capacity.

As the dollar has depreciated from its mid-1985 level, the cost differential between the U.S. producers and their leading foreign rivals has fallen. However, over one-third of foreign costs are dollar denominated. Thus, the yen would have had to appreciate more than 40 percent—from 240 to less than 150 to the dollar—to equalize U.S. and Japanese operating costs at the efficient integrated plants illustrated in table 3-4. And West German integrated producers continue to have an operating-cost advantage over their U.S. rivals with the deutsche mark at 2.20 per dollar, as the following variations in exchange rates suggest:[14]

14. Based on costs in table 3-4.

Type of cost	U.S. costs (dollars)	West German costs (dollars) at a DM-dollar rate of				Japanese costs (dollars) at a yen-dollar rate of			
		2.00	2.50	2.90	3.00	150	200	240	250
Operating	403	409	354	324	317	386	319	286	280
Financial	43	58	46	40	39	98	73	61	59
Total	446	467	400	364	356	484	392	347	339

The higher financial costs per ton in the newer Japanese plants offset some of their advantage in operating costs. However, to sell in the U.S. market, exporters must incur transportation costs and import duties. Therefore, Japanese firms probably would not be fully cost competitive with the best U.S. plants in the U.S. market with the value of the yen above 200 per dollar, nor would German plants with the value of the deutsche mark above 2.50 per dollar.

Plant Closures

Between 1977 and 1985 the major integrated companies in the United States reduced their capacity from about 150 million tons of raw steel to about 110 million tons. Much of this reduction has been attributed to the lack of growth in demand and to import competition. An inspection of the facilities closed by the major producers provides a somewhat different explanation—namely, that minimill competition is an important contributor to the major companies' problems.

It is difficult to characterize and measure facility closings in the steel industry because of the number of different production facilities in a single plant or complex. Steel companies rarely close a plant abruptly. Rather, they retire individual facilities or mills in the plant after allowing them to wear down gradually for a number of years. Eventually, the entire plant may be closed, but most of the retirements in the past decade have come from partial plant closures.

Table 3-5, which lists the most important facility closures announced by the major companies, reveals a number of interesting phenomena. First, most of the closures have been in plants producing bars and wire rods—the two products dominated by minimills today. Second, while the major companies have built electric furnaces to save declining plants (at Johnstown and Steelton for Bethlehem Steel; Pittsburgh for LTV Steel Company), there have been several sales or closures of electric-furnace plants—LTV's Pittsburgh plant, Bethlehem Steel's Los Angeles

Table 3-5. *Plant Closures by Integrated Steel Firms in the United States, 1977–85*

Firm	Plant location	Facilities or products affected
Armco	Houston, Texas	Bars, tubes, plates
	Sand Springs, Oklahoma (sold)	Bars
	Marion, Ohio (sold)	Bars
Bethlehem Steel Corp.	Baltimore, Maryland (partial)	Wire rods, pipes
	Bethlehem, Pennsylvania (partial)	Bars, structural shapes
	Lackawanna, New York (partial)	Rails, bars, structural shapes, sheets
	Johnstown, Pennsylvania (partial)	Bars, plates
	Los Angeles, California (sold)	Bars, wire rods
CF&I Steel Corp.	Pueblo, Colorado (partial)	All basic oxygen furnaces
Inland Steel Co.	Indiana Harbor, Indiana (partial)	Open hearth furnaces, bar mills
Kaiser Steel Corp.	Fontana, California (reopened as California Steel)	Sheets, plates, structural shapes
LTV Steel Co.	Aliquippa, Pennsylvania	Sheets, bars
	Buffalo, New York	Bars
	Chicago, Illinois (partial)	Basic oxygen furnaces, wire rods
	Pittsburgh, Pennsylvania	Electric furnaces, bars
National Steel Corp.	Buffalo, New York	Blast furnaces
	Great Lakes, Michigan (partial)	Basic oxygen furnaces
	Liberty, Texas	Pipes and tubes
United States Steel Corp.	Chicago, Illinois (partial)	Plates, wire rods, structural shapes
	Fairfield, Alabama (partial)	Structurals, plates, galvanizing line
	Gary, Indiana (partial)	Bars, rails
	Cuyahoga, Ohio	Wire rods, cold-rolled strip
	Fairless Hills, Pennsylvania (partial)	Wire rods, wire, bars
	Provo, Utah (partial)	Structural shapes
	Monongahela Valley, Pennsylvania (partial)	Basic oxygen furnaces, bars, structurals, tubes
	Youngstown, Ohio	Bars, sheets
Wheeling-Pittsburgh Steel Corp.	Allenport, Pennsylvania	Tubes, sheets
	Benwood, West Virginia	Pipes
	Monessen, Pennsylvania (partial)	Bars
Wisconsin Steel	Chicago, Illinois	Bars

Sources: William Hogan, *U.S. Steel Industry in the 1980s* (Heath Lexington, 1984); annual reports of the companies; and AISI, "Shutdown of Steel Plants and Producing Facilities" (Washington, D.C.: AISI, June 1, 1986).

and Seattle plants, and Armco's plants at Sand Springs, Oklahoma, Marion, Ohio, and Houston.[15] Third, many companies have reduced their capacity for making raw steel but kept the associated rolling mills. These plants are candidates for imported semifinished steel or even steel supplied by other U.S. producers, including minimills.

The integrated steel companies, rather than keeping obsolete raw-steel works open or building new coke ovens, blast furnaces, basic oxygen furnaces, and continuous casters, appear to be willing increasingly to rely on other producers to feed their raw steel requirements for a narrower and narrower line of finished steel products. Small electric furnace companies may very well offer to meet some of this demand.

In only a few cases have the integrated firms' closures involved flat-rolled products, and even these closures did not account for a very large share of the industry's capacity. The total flat-rolled steel capacity has not shrunk appreciably. On the other hand, the integrated firms' bar and wire rod capacities have been decimated. The reason for the difference is obvious; minimills compete in the latter product markets, but not in the former.

Continuing Imbalances

As a result of the ambitious investment programs of the mid-1970s that were cut short by the decline in demand, the integrated firms are left with a large number of plants that are badly out of balance. Many are short of coking capacity.[16] Some do not have enough raw-steel capacity to feed their rolling mills. Many have far too little continuous casting. Most have one or more facilities that should be modernized or replaced.

The status of the major companies' steel plants, summarized in table 3-6, does not reflect the problems they may have with individual blast furnaces, coke ovens, or rolling mills. Rather, the tabulation simply demonstrates how a decade of dashed expectations has left most of the integrated companies' plants out of balance. The problems are so severe

15. The Sand Springs plant was sold to Sheffield Steel. The Bethlehem plant in Seattle is now Seattle Steel.

16. Apparently, many integrated companies are planning to rely on abundant supplies of coke in world markets for the indeterminate future. This is a complete reversal of the shortage psychology of just ten years ago. The industry may be anticipating the development of new technologies that allow coal to be injected directly into blast or steel-making furnaces.

Table 3-6. *Imbalances at Major Integrated Steel Plants in the United States, 1985*

Company and plant location	Raw steel capacity (millions of tons)	Type of furnace	Imbalances	Expected status, mid-1990s[a]
Armco				
Middletown, Ohio	3.3	Basic oxygen	Short of blast	Likely to survive
	1.2	Open hearth	furnace and coking capacity	
Ashland, Kentucky	2.0	Basic oxygen	Small-capacity; older hot-strip mill	Probably will close hot-strip mill and ship slabs to Middletown
Butler, Pennsylvania	1.0	Electric	None	Will survive
Kansas City, Missouri	1.5	Electric	Limited continuous casting	Likely to suffer from minimill competition; uncertain
Bethlehem Steel Corp.				
Bethlehem, Pennsylvania	2.0	Basic oxygen	None	May survive
Burns Harbor, Indiana	5.3	Basic oxygen	Short of coke; building second caster	Certain to survive
Johnstown, Pennsylvania	1.4	Electric	No caster	Unlikely to withstand minimill competition
Lackawanna, New York	0.0	None	A fragment of a plant	Coke capacity will probably survive
Sparrows Point,	3.3	Basic oxygen	Building	Likely to remain,
Maryland	3.0	Open hearth	first caster	perhaps as seller of slabs
Steelton, Pennsylvania	1.4	Electric	None	Rail mill may remain but bar mills subject to minimill competition
Inland Steel Co.				
East Chicago, Indiana	6.3	Basic oxygen	Short on coke;	Will survive
	2.5	Open hearth	needs additional	
	0.5	Electric	continuous casting	
LTV Steel Co.				
Warren, Ohio	2.0	Basic oxygen	Short on coke and blast furnace capacity	Uncertain
Canton, Ohio	1.0	Electric	Limited finishing capacity	Uncertain
Cleveland, Ohio	6.3	Basic oxygen	Short on coke; needs second continuous caster	Probably will survive
Indiana Harbor, Indiana	3.0	Basic oxygen	Short on coke	Will survive
Chicago, Illinois	1.0	Electric	No continuous caster	Older bar mills may close; electric furnaces may survive
Gadsden, Alabama	1.5	Basic oxygen	Short on coke; no continuous caster	Must be divested; likely to close raw steel-making operation

Table 3-6 *(continued)*

Company and plant location	Raw steel capacity (millions of tons)	Type of furnace	Imbalances	Expected status, mid-1990s[a]
National Steel Corp.				
Granite City, Illinois	2.2	Basic oxygen	Short on coke; poor blast furnaces	Likely to survive
Detroit, Michigan	2.7 0.8	Basic oxygen Electric	Excess rolling capacity; could import slabs	Likely to survive
Rouge Steel Co.				
Dearborn, Michigan	3.6	Basic oxygen, electric	Needs additional continuous casting	Likely to survive
United States Steel				
Clairton, Pennsylvania	0.0	None	A coking facility	Likely to survive
Monongahela Valley, Pennsylvania	2.9	Basic oxygen	No continuous casting	Likely to close
Fairfield, Alabama	2.9	Basic oxygen	Short on coke; needs a continuous caster	Likely to survive
Fairless Hills, Pennsylvania	3.6	Open hearth	Poor raw steel facilities; old hot-strip mill; no slab caster	Unlikely to survive
Provo, Utah	3.0	Open hearth	Old open hearths; no continuous caster; poor finishing facilities	Unlikely to survive
Pittsburgh, California	0.0	None	Only rolling mills	Likely to survive
Gary, Indiana	7.0	Basic oxygen	Needs second continuous caster	Will survive
Lorain, Ohio	2.8	Basic oxygen	Needs additional continuous casting	Will survive
Baytown, Texas	1.2	Electric	None	May survive
South Chicago, Illinois	0.7	Electric	A fragment of a plant	Uncertain
Weirton Steel				
Weirton, West Virginia	4.0	Basic oxygen	Needs additional continuous casting	Likely to survive
Wheeling-Pittsburgh Steel Corp.				
Monessen, Pennsylvania	1.5	Basic oxygen	Poor raw steel facilities	Uncertain
Steubenville, Ohio	3.0	Basic oxygen	Poor finishing facilities	Likely to survive

Source: Richard Serjeantson, Raymond Cordero, and Henry Cooke, eds., *Iron and Steel Works of the World*, 8th ed. (Surrey, England: Metal Bulletin Books, 1983).

a. Authors' predictions.

that, given demand forecasts, a majority of the integrated plants have little prospect of remaining in operation into the 1990s (see chapter 7). Integrated companies cannot justify rebuilding coke ovens, blast furnaces, and steel-making furnaces at plants such as Gadsden, Lackawanna, Fairless, Geneva, and Warren. Electric furnace plants without casters and with older rolling mills cannot easily be modernized to compete with current or prospective minimills. In a few cases, investments in continuous casters will improve the efficiency of older plants, but such investments will not be profitable for many plants.

Integrated steel facilities in developed countries face difficult times ahead, and only the most efficient will survive. Most of the Japanese plants are well balanced and modern, and so are a few European plants. But few U.S. integrated plants can be said to be of "world class."[17] As a result, many plants will wither away in the face of competitive assaults from steel producers in developing countries and from U.S. minimills.

The Political Equation

Having partially renovated or expanded many plants, the integrated companies find themselves in a ticklish situation. They desperately need the political support of their workers and the residents of the communities in which they have facilities if they are to obtain protection from foreign competition. Obviously, this makes it difficult to close plants or individual facilities within those plants. While many facilities have been closed, a number of inefficient plants remain, absorbing scarce investment capital. The results of such investments are obviously extremely poor, causing the external capital markets to be very hesitant to underwrite new ventures by these managements.

Cost of New Facilities

There can be no doubt that minimills can build steel plants at a much lower cost per ton of capacity than integrated firms. For instance, it would cost about $1,400 per ton of capacity to build a new integrated plant producing cold-rolled coil; more than $600 of that is for raw steel and casting facilities (table 3-7). On the other hand, minimills producing

17. See John Tumazos, "Metals and Mining Industries" (New York: Oppenheimer and Co., September 29, 1983). Tumazos estimates that only 58 million tons of U.S. integrated capacity is "world class."

Table 3-7. *Cost of Constructing an Integrated Plant to Produce Cold-Rolled Coil, 1985*

Process	Cost per ton of capacity (dollars)	Input per ton of finished product (tons)	Cost per ton of finished product (dollars)
Sintering	105	0.33	35
Coke oven	380	0.44	167
Blast furnace	240	0.91	218
Basic oxygen furnace	105	1.09	114
Continuous caster	105	1.09	114
Hot-strip mill	260	1.05	273
Continuous cold finishing[a]	500	1.00	500
Total	. . .	. . .	1,421

Source: Authors' estimates.
a. Pickling and cold rolling, annealing, tempering, and finishing.

wire rods have generally been built at a cost of $250 to $300 per ton of capacity.

The implications of the high cost of a new integrated mill are quite clear. The total cost of producing a ton of cold-rolled coil in an efficient U.S. integrated plant in 1985 was $446, of which $403 was operating costs (table 3-8). Though the operating costs in a new plant are only $319 per ton, the savings are swallowed up by capital costs (depreciation, interest, and taxes), which soar to $176 per ton (based on the construction cost of $1,421 per ton shown in table 3-7 and following conventional accounting procedures). Total production costs for this new plant are therefore about $495 under conventional accounting. If, however, the real cost of capital is assumed to remain at 7 percent over a plant life of fifteen years, capital costs would be only $153 per ton. This would suggest that a new plant could operate at total costs of $472, or about 17 percent more than the operating costs of the more efficient of the current U.S. integrated plants. Unlike the minimills, integrated firms simply cannot reduce costs by building new plants.[18] It is for this reason that no new plant has been built since the 1960s.

It is even likely that integrated companies could not build small electric furnace plants at a cost similar to those achieved by the smaller companies. Indeed, Armco and Bethlehem Steel have been moving away from such facilities, not building them. Each of the major integrated

18. This is not to say that some new cold-finishing complexes may not be built economically. Indeed, Inland Steel Co. is considering the possibility of a new rolling facility to finish steel produced in existing basic oxygen furnaces and hot-strip mills.

Table 3-8. *Cost of Producing Cold-Rolled Coil in Existing and New Integrated Plants in the United States, 1985*
Dollars per ton of finished product

Item	Existing plant	New plant
Operating costs	403	319
Labor	129	87
Iron ore	67	55
Scrap	18	19
Coal or coke	50	38
Other energy	24	12
Miscellaneous	115	108
Depreciation	24	95
Interest	12	71
Taxes	7	10
Total	446	495

Sources: Tables A-4, A-5.

firms has a large engineering staff that has been accustomed to engineering large complexes with equipment with a useful economic life that stretches into decades, not just a few years. Some minimills are being built as turnkey operations by outside contractors who deliver them to the purchasers as complete operating facilities. The integrated companies have been reluctant to allow other engineering companies to design plants for them. Nor would the United Steelworkers of America take kindly to decisions by the integrated companies to abandon their integrated facilities in favor of minimills that are far less labor intensive. Workers at the declining integrated plants have very lucrative early retirement options that would further frustrate such a move. Finally, the integrated firms have not been at the forefront of technical change in the operation of electric furnaces or in continuous casting of small-diameter products.

Cost-cutting Efforts

Continuing import competition, facility imbalances, sluggish market growth, and the continued growth of minimills have forced the integrated producers to begin to restructure operations, close noncompetitive facilities, and reduce variable costs. Contracts for raw materials and labor have been renegotiated. Between 1981 and 1985, real unit costs of integrated production fell by about 15 percent. Further labor cost reductions and facility restructuring could reduce costs another 5 percent in 1986.

European and Japanese producers are also responding to the weak world steel market by reducing costs. Therefore the U.S. companies are likely to be faced with further reductions in real world steel prices, and, unless the dollar falls substantially from its mid-1986 value, it is unlikely that the U.S. integrated companies will be able to recapture much of the market lost to imports. Nor can the integrated companies hope to be competitive with the minimills, which are also steadily reducing their costs of producing bars, structural shapes, and wire rods.

Summary

It is unlikely that United States Steel, Bethlehem Steel, Inland Steel, LTV, or even Armco can break out of their mold as large-scale, fully integrated producers. They are saddled with plants employing older technology, built under assumptions about prices and demand growth that have proved to be incorrect. They must now manage these assets and husband their scarce capital to modernize and renew the plants that have some useful life left. They cannot walk away from these plants and build a new set of smaller, efficient electric furnace plants for a major share of their output. Their workers, bankers, and political allies simply would not allow it.

IV

Changing Technology and the Minimills

It is clear from chapter 2 that minimills have been very successful in adopting new technology. By improving their electric furnaces, continuous casting, and rolling mills and by adopting ladle metallurgy, they have significantly improved performance, lowered input requirements (especially for labor), and reduced unit capital costs.

Further changes in technology now in prospect indicate that the sector will continue to grow. Improvements in a number of processes should make possible reductions in the real cost of the products they now specialize in, and completely new technology should allow the minimills to expand rapidly into more sophisticated lines.

Recent Technological Changes

Over the last few years there have been remarkable changes in the simple technology employed by minimills. Electric furnaces and casters have been in general use since the 1960s, but until recently most steel billets were made by first casting ingots and then rolling them into billets. The secret of the success of the modern minimill can be attributed, with some oversimplification, to improvements in electric furnace steel making and the billet caster.

Steel Making

In the early 1960s the electric furnace was essentially used for the slow refining of sophisticated products like alloys and stainless steels. However, technological improvements since then have made it possible to reduce the heat times in electric furnaces. While tap-to-tap times of about ten hours and 0.15 kilovolt amperes of electricity per ton of vessel capacity were the norm in the early 1960s, ten years later the time from beginning to end of the process had been halved and the voltages the furnaces could handle had been doubled. By the 1980s, tap-to-tap times had fallen to between one and two hours, and the electric furnaces could accommodate electric charges of about 0.75 kilovolt amperes per ton to melt scrap. The huge reduction in the amount of time the metal was in the furnace translated into less labor required, less wear on electrodes, and less power consumed per ton of output. The following performance measures are representative of the advances in the minimill sector that improvements in electric furnaces over two decades have made possible:

Electric furnace performance	1965	1975	1985
Tap-to-tap time (hours)	10.00	5.00	1.50
Electrode use (pounds per ton)	14.00	12.00	9.00
Electricity use (kilowatt hours per ton)	550.00	525.00	520.00
Electric charge (kilovolt amperes per ton of vessel capacity)	0.15	0.30	0.75
Labor use (man-hours per ton in steel making, including casting)	2.90	1.75	0.90

It is interesting to note that two different approaches to electric furnace production of carbon steel began to develop in the 1960s and 1970s. One approach, essentially that of the large integrated producers who use some electric furnaces, emphasized larger and larger vessels (up to 400 tons per heat, or batch) at relatively low ratios of voltage to vessel capacity to get maximum productivity. The alternative approach, employed by the very small minimills, was to use small vessels (35 tons to 50 tons per heat). Both approaches required at least two furnaces because the vessels frequently underwent a lengthy relining procedure. During the 1960s and 1970s when the move to high power began, the large vessel had an advantage because of its greater scale.

Higher power reduced tap-to-tap times and meant larger and larger tonnages from each vessel. However, this created two problems. By the late 1970s the large vessels created such large power requirements that the attendant demand surges either caused brown-outs on the utility grid or the firm had to invest in its own electricity generating station. Furthermore, the vessels often held too much steel to pour into the ladles and tundishes that serve as reservoirs for continuous casting. When the electric furnace had to act as a holding facility, the improvements in tap-to-tap time were dissipated. Finally, because caster pouring schedules are very precise, the benefits of performance improvements shifted to smaller vessels that could use the greater electrical charge and match their output with the casting rate.

As improvements in refractories have lengthened the time between relinings of furnaces and the speed of replacement, the need for two vessels has been reduced. Oxy-fuel burners that supplement the electrical charge have shortened heat times in the electric furnace and reduced the consumption of electricity. Similarly, improvements in the casters' performance have reduced the need for multiple casters served by multiple electric furnaces. As a result, in the late 1970s, the optimal configuration became a single medium-sized electric furnace, with oxy-fuel burners, feeding one continuous caster.

The single 120-ton electric furnace installed by Raritan River Steel Company, with three oxy-fuel burners, is regarded as the most efficient way to produce steel with an electric furnace at current energy prices. Other relatively minor improvements include water cooling of panels and roofs, use of the foamy slag that floats on molten steel to reduce wear on the electrodes above the steel, and further refining in the ladle outside the steel furnaces. With these changes, minimills have been able to reduce their steel-making costs relative to their integrated competitors' costs. And many of the changes have allowed minimills to begin to produce high-grade carbon steel for wire rod and seamless tube, which was not generally possible ten years ago.

Billet Casting

One of the major changes in technology has been the development of billet casting. For long products—rods, bars, structural shapes, and rails—this technology essentially replaced three steps (ingot casting, the rolling of the ingots into blooms, and the rolling of blooms into billets) with a single process. For flat products, such as sheets and plates, continuous casting replaced ingot casting and primary rolling with one process. The change in process has brought dramatic improvements in energy efficiency, product yields, labor productivity, and capital construction costs. The yields from scrap to finished bar are 18 percentage points higher in a minimill with continuous casting than in one that uses the ingot process (84.5 percent versus 66.5 percent).

The introduction of continuous casting in the production of billets has reduced labor requirements by about 65 percent—from 1.3 man-hours per ton with ingot casting to about 0.45 hours. The cost of building electric furnaces and continuous casters (in 1985 dollars) is about $140 per ton of finished billets while the older ingot-casting facilities would have cost about $280 per ton. Whether or not the full costs of a continuous billet caster are less than the incremental costs of a traditional ingot process, it is clear that all new plants will use the continuous casting process.

Aside from the major change brought about by the use of continuously cast billets, changes in the billet caster itself have made it a more attractive facility. Changes in the shape of the mold to speed the casting process, better cooling procedures, larger tundishes, and faster adjustment of mold sizes have all resulted in significantly lower man-hours,[1] a better quality billet (fewer rejects), and, perhaps most important, a lower cost to construct a plant producing billets.

Hot Rolling

The technical changes have extended to the rolling of steel products themselves. The bars, rods, and light shapes that minimills specialize in have all enjoyed remarkable changes in technology, with faster rolling speeds, better yields, lower energy usage, lower labor requirements, better product quality, and lower construction costs. Perhaps no facility exhibits more graphically the major changes that have occurred than the rod mill, as these performance characteristics of a typical one suggest:

Rod mill performance	1972	1985
Rolling speed (feet per minute)	10,000	20,000
Yield (tons of bars per ton of billets)	0.90	0.96
Labor use (man-hours per ton)	1.65	0.85
Electricity use (kilowatt hours per ton)	180	160
Other energy use (millions of Btus per ton)	3.00	2.25

The remarkable improvement in efficiency is the result of numerous technical changes. The Morgan no-twist mill, for example, rolls wire rods at a much greater speed than its predecessors. The Stelmor coil layer significantly increases the speed at which the rod mill can operate by looping the wire rod as it emerges from the rod mill. Other technological advances have brought progress in rolling all of the long products.

1. For example, labor usage at the average billet caster in the United States decreased from 0.80 man-hours per ton in 1972 to 0.45 in 1980, and to 0.30 in 1985.

Anticipated Technical Changes

Future technical changes are likely to be far more revolutionary than those of the recent past. Changes can be foreseen in melting, casting, and hot-rolling processes.

Steel Making

One of the major contributors to the improved performance of electric furnaces has been the use of higher electric power levels. Many mills fall far short of the current optimal ratio of 0.75 kilovolt amperes per ton of vessel capacity. Most existing furnaces will be replaced or retrofitted with transformers that allow their ratios to equal those in new steel furnaces, and oxy-fuel burners will be installed to substitute less costly fuel for electricity. This of course does not signal the end of the search for still better processes of steel making, but rather a change in direction. The major change in the steel furnace itself that seems likely will be the replacement of the traditional top-poured vessel with an eccentric bottom-poured vessel. The new vessel has a spout approximately one-third of the way from the top of the furnace that allows quicker pouring of steel with lower oxidation and fewer impurities. This method increases the life of the lining at the lip of the vessel and allows the use of water-cooled panels further down the vessel sides, thus reducing the use of refractories. It also reduces the consumption of electricity by about 4 percent, electrode use by one-fourth, and labor requirements by one-third because of faster tap-to-tap times. Electrode use is further lowered if foamy slag is incorporated in the steel-making process.

The eccentric bottom-poured furnace increases capital costs, but those costs are more than offset by the increases in output brought about by the faster heat times (and faster pouring times). Raritan River and North Star began installing this type of furnace in 1985. And a West German mill builder, Mannesmann Demag Corporation, claimed that by 1988 these furnaces would be the only new electric furnaces built. Retrofitting with this type of facility requires about $10 million for one vessel with 150 tons of capacity, and the realized savings are estimated at about $10 per ton in operating costs. Therefore, a mill producing 600,000 tons of steel annually could save $6 million, surely a satisfactory annual return on an investment of $10 million.

Another approach to improved electric furnace steel making is the use of direct rather than alternating current in the furnace, requiring a transformer (costing about $4 million in 1985 dollars) and a single electrode (instead of three), which allows the current to arc between the electrode and the bottom of the furnace.[2] Electrode use is believed to be reduced by about 50 percent. Wear of the refractories on the side walls is lowered (because there are fewer electrodes), but refractory wear on the furnace bottom is increased. This approach is believed best for rebuilding small furnaces.

Still another improvement expected in electric furnace steel making is the continuous feeding of preheated scrap. Nucor intends to do this by feeding heated scrap on an enclosed conveyor through the side of an electric furnace. Gas by-products from the furnace, piped into the conveyor, raise the temperature of the scrap to about 900 degrees Fahrenheit. This continuous feeding of heated metal will reduce both peak electricity requirements (and hence demand changes) and total requirements per ton. Savings in refractories are also expected. The biggest savings anticipated, however, are through increases in the output produced by a given furnace capacity, as continuous charging permits virtually continuous tapping. Frequent pouring of molten steel leaves a heel of molten steel in the furnace that itself makes possible continuous arcing of electricity, reducing electricity consumption. The heel also reduces wear on electrodes and refractories, especially if direct current is employed.

Many small two-furnace shops are expected to convert one of their furnaces to refining; the remaining furnace would continue to be used for melting, with no loss of volume. Continuous charging alone would require an investment of $3.75 per annual ton of capacity but might yield $5 per ton savings in operating costs. Retrofitting for continuous charging and tapping would cost up to $40 per annual ton of capacity but save up to $20 per ton in operating costs.

Ladle metallurgy, a refining method that improves product quality, is widely used in minimills, but only on a small scale. Much greater reliance on this technique, which keeps steel at high temperatures during refining, is expected in the near future. Refining in the ladle reduces the time steel must remain in the furnace, reducing tap-to-tap times by up to 15 percent.

2. A discussion of direct current technology can be found in George J. McManus, "Nucor Plugs in Its Direct Current Furnace," *Iron Age,* vol. 228 (October 4, 1985), pp. 50–61.

Some experts believe that the major technical change of the next few years will be in ladle metallurgy. Operating cost savings of up to $20 a ton are expected with minimal capital costs.

Continuous Casting

No major changes are anticipated in the method of casting billets. Horizontal casting shows promise of faster casting time and better quality than the current curved casters. This technology is difficult to control and is likely to be developed slowly. A more immediate prospect is the development of beam-blank casting—the casting of a bloom in an H shape, which requires much less rolling than a conventionally cast bloom. Nucor has recently announced plans to build a new structural mill that will utilize this process to produce shapes up to twenty-four inches in cross section. Some minimills, such as Chaparral and Bayou, are already producing medium-sized shapes.

Another major change in casting in the next few years is expected to be in casting thin slabs for rolling into sheets. One of two fundamentally different approaches being considered—which will take about ten years to be implemented—would cast sheets of approximately 0.10 inch directly from molten steel. Both United States Steel and Bethlehem Steel have received government research funds to investigate this technology.[3] Not only would it radically change the nature of casting, but more fundamentally, it would completely replace today's hot-strip mill with a much smaller mill. Aside from technical difficulties, this approach raises serious questions of whether its product will be satisfactory for most applications because the grain configuration of a cast sheet is quite different from that of a rolled sheet.

The alternative approach involves casting 1.5-inch slabs (of various widths) from molten steel. The slabs would be about one-fifth the thickness of today's 8-inch slabs and could possibly shorten the rolling process from approximately four roughing stands and as many as six finishing stands to one roughing and four finishing stands. This technology is much closer to implementation than the process that moves directly from molten steel to cast sheets. Construction has begun on a pilot facility at Nucor Corporation's Darlington, South Carolina, plant.

3. "Westinghouse, Armco Working to Develop Continuous Caster for Thin Strip," *Iron Age*, vol. 228 (April 5, 1985), p. 18.

This technology is expected to be competitive at about 500,000 tons per year, but product quality is expected to be below average. Nucor plans to supply commercial grade sheet—suitable for barrels, drums, and steel decking—allocating approximately 25 percent of its output to its own fabricating business.

Hot Rolling

No major technical changes are anticipated in the typical bar and rod mill. However, increases in the speed of rolling bar products can be expected as more of them are rolled on rod mills—in coils rather than in lengths, which reduce rolling speeds. The technical changes noted above will make it possible to reduce the semifinished shapes more quickly, with fewer passes.

The most significant technical change in hot rolling is expected to be the development of the mini hot-strip mill, making possible smaller-scale low-cost production of sheets and plates. The current approach to rolling sheet steel from slabs involves continuous or semicontinuous rolling to reduce, say, an 8-to-10-inch slab to a sheet 0.1 inch or less in thickness. The minimum efficient scale of these hot-strip mills is over 3 million tons and they cost over $450 million to build. The size of the market they must serve and the sizable capital costs involved have prompted efforts to find a lower-cost alternative. The Steckel mill is the traditional alternative, using a reversing rougher and a reversing finishing stand to reduce slab to sheet (as opposed to four roughing stands or one reversing rougher, plus as many as six finishing stands). The Steckel approach, however, suffers from high yield losses, poor energy efficiency, and poor product quality.

The major potential for lower-cost production of sheet, short of sheet casting, would appear to rest on simplifying rolling technology by reducing the number of rolling stands. A thin slab caster that can produce, say, a 1.5-inch slab can achieve the same results as the roughing stands in a conventional hot-strip mill, realizing significant capital and operating savings, but it does not appreciably reduce the energy required to produce a sheet. About 90 percent of the energy required to convert a 10-inch slab to a 0.1-inch sheet is expended in reducing the product from 1.5 inches to 0.1 inch. Heat, which is crucial for rapid rolling and for a metallurgically sound product, is lost at a rate that is proportional to the ratio of surface to volume, and below 1.5 inches in thickness, the heat loss of a slab is exceptionally rapid. To solve that problem, the continuous hot-strip mill uses massive power and speed to finish rolling when the product is hot, while the Steckel mill relies on retaining heat (in coil boxes) during the slower, reversing rolling.

The aim in developing a mini hot-strip-mill technology is to achieve the operating efficiency and product quality of the continuous hot-strip mill, but with fewer rolling stands. A technology that would shorten the process where energy losses and power demands are greatest would offer even greater capital cost savings than the thin slab caster. A major step in that direction—a modern application of the Steckel approach—has been spearheaded by Tippins. It has installed mini hot-strip mills to produce plate and thick sheet in South Africa and Canada and to produce coiled plate at Tuscaloosa, Alabama. According to Tippins, this type of mill has very low capital costs ($75 million for capacity over 500,000 annual tons), good yields (95 percent), and excellent energy efficiency, and it produces a quality product. The mill will take slabs varying from 10 inches to 1.5 inches in thickness and roll them in widths up to 120 inches and to thicknesses of 0.1 inch or less.[4]

For certain domestic fabrication markets, small domestic companies with mini hot-strip mills may be lower-cost sources of steel than either imports or large domestic producers. A mini hot-strip mill could initially use purchased slab (as Tuscaloosa does) and later add on an electric furnace and slab caster. If it were linked with thin slab casting, there could be additional cost savings both in steel making and in rolling.

Thin slab casting and the mini hot-strip mill could revolutionize the steel business, making it possible to build small-scale steel plants to supply commercial grade sheets and plates. Eventually the mills would be upgraded to produce high-quality hot- and cold-rolled sheet. Because minimills' capital costs are much lower than those of large-scale integrated mills, more rapid modernization of the industry and a more market-oriented approach to production would be possible. Small-scale facilities can fit into small areas and narrow product ranges. When these advantages are combined with more sophisticated ladle metallurgy and better scrap sorting and testing, minimills will begin to enter the market for flat products. Though the quality of steel produced at traditional integrated mills may not be matched for some time, the refinement of minimills' basic technologies will bring rapid improvements.

4. Figures are from industry sources. In June 1986 United States Steel agreed to supply Tuscaloosa with slabs from its Birmingham plant.

Technological Change and Production Costs

Nominal production costs in both minimills and integrated mills have fallen in the 1980s. But the costs of production for bars and rods in integrated facilities have not fallen relative to those of minimills. Integrated firms have been unable to match minimills' relative cost improvements; however, reorganizations and renegotiations of materials and labor contracts in 1986 will lower production costs at some integrated plants.

Bars and Wire Rods

The cost of building a state-of-the-art minimill to produce 750,000 tons of wire rod per year is approximately $300 per ton of finished product capacity, as the following breakdown of construction costs, in dollars per ton, shows:[5]

| | Capital | Cumulative |
Process	cost	capital cost
Electric furnace	100	100
Casting	40	145
Rod mill	155	300

This new minimill offers substantial improvements in performance over plants now in existence—lower electricity consumption, higher yield from raw steel to finished products, and lower labor usage (table 4-1). Operating costs per ton are approximately $16 lower than those of the best of the existing plants. But the new plant requires an increase in capital costs of about $9. Its unit costs are above the incremental costs but below the unit costs of production at the older plants. Obviously, given the fact that costs at integrated plants are higher, the minimill advantage over integrated producers continues. This suggests that for most bar, rod, and small structural products, the minimills will continue to take markets from the integrated companies, forcing mill closures at integrated firms.

5. The cumulative capital cost for casting is equal to $100 per ton for the electric furnace and $40 per ton for the caster, for a total of $140 per ton of billets. This figure must be multiplied by 1.035 to reflect the yield loss between the billet and the finished rod. Therefore the cumulative capital cost through the casting process equals $140 × 1.035, or $145.

Table 4-1. *Cost of Producing Wire Rod in Efficient Existing and Prospective Minimills with 250,000 Tons Annual Capacity, 1985*

Item	Existing mill	Prospective mill
	Dollars per ton of finished product	
Operating costs	224.00	208.00
Labor	34.00	27.00
Scrap	92.00	92.00
Electricity	30.00	27.00
Other energy	12.00	11.00
Electrodes and refractories	18.00	14.00
Miscellaneous	38.00	37.00
Depreciation	12.00	22.00
Interest	18.00	17.00
Taxes	2.00	2.00
Total cost	256.00	249.00
Addendum		
Input prices		
Scrap (dollars per ton)	85.00	85.00
Electrodes (dollars per pound)	1.20	1.20
Electricity (dollars per kilowatt hour)	0.045	0.045
Labor (dollars per man-hour)	17.50	17.50
Efficiency measures		
Man-hours per ton	1.95	1.55
Electricity use (kilowatt hours per ton)[a]	485.00	435.00
Electrode use (pounds per ton)	7	5
Yield to finished product (percent)	96	97

Sources: Tables A-2, A-3; data provided to the authors by various producers. Tons in all tables are net tons unless otherwise noted.

a. Electric furnace and caster only.

Sheet Production

Minimills are now at the threshold of a major breakthrough into sheet production. The large integrated companies have dominated the flat-rolled markets because of the enormous scale economies in rolling slabs into sheets or plates. The rising costs of integrated steel production and impending technological developments in mini hot-strip mills and in slab casting will see this dominance challenged in the next few years.

The costs of building a new integrated plant to produce the typical product, cold-rolled sheet in coils (see chapter 3), are such that they cannot be justified at current or prospective prices of cold-rolled sheet. Moreover, they are above the costs at the more efficient current plants

Table 4-2. *Cost of Constructing a New Minimill Producing Cold-Rolled Coil with Conventional Continuous Casting, 1985*
Dollars per ton of finished product[a]

Item	Cost	Cumulative cost
Electric furnace	112	112
Continuous caster	90	202
Hot-strip mill	202	404
Cold finishing mills[b]	300	704

Source: Data provided to the authors by various steel engineers and producers.

a. Based on 90 percent utilization of production capacity.

b. Comprised of pickling facilities ($65), cold-rolling tandem mill ($135), annealing mill ($70), and tempering mill ($30).

because of the enormous capital charges required for a new integrated plant.

A new minimill, on the other hand, that could produce cold-rolled coil with conventional casting techniques—continuous casting of eight-inch slabs—would cost about $700 per ton of finished capacity to construct (table 4-2), or about 50 percent of the cost of an integrated plant. The costs of producing cold-rolled coil in a new minimill plant with about 500,000 tons per year of capacity would be at least 10 percent lower than in the current integrated plants and almost 20 percent lower than in a new integrated plant (table 4-3). However, the minimill product would be of somewhat lower quality than the sheet produced in an integrated plant because of impurities in the scrap and the physical properties imparted by the less expensive hot-rolling mill. The initial market for minimill sheet would therefore be for lower-quality applications, not for the exterior panels on automobiles or appliances.

As improvements are made in thin slab casting and mini hot-strip mills, the cost of a minimill operation will continue to be below the costs of a typical integrated producer. The problem, however, will be product quality, and there will be room for only a few lower-quality mills in particular market niches, regardless of the cost of production. The success of any minimill producing flat products will turn on its ability to deliver a higher-quality product than its minimill competitors.

Expansion and the Sources of Capital

With nominal costs of production still falling, the minimill sector has the potential to continue growing for the rest of this century unless there

Table 4-3. *Cost of Producing Cold-Rolled Coil in Existing and New Integrated Plants and Prospective Minimills, 1985*
Dollars per ton of finished product[a]

Item	Integrated Plant		Prospective minimill
	Existing	New	
Operating costs	403.00	319.00	311.00
Labor	129.00	87.00	65.00
Iron ore	67.00	55.00	. . .
Scrap	18.00	19.00	96.00
Coal	50.00	38.00	. . .
Other energy	24.00	12.00	56.00
Miscellaneous	115.00	108.00	94.00
Depreciation	24.00	95.00	47.00
Interest	12.00	71.00	35.00
Taxes	7.00	10.00	5.00
Total costs	446.00	495.00	398.00
Addendum			
Input prices			
Iron ore (dollars per ton)	40.00	40.00	. . .
Scrap (dollars per ton)	80.00	80.00	85.00
Labor (dollars per man-hour)	22.50	22.50	17.50
Electricity (dollars per kilowatt hour)	0.045	0.045	0.045
Coal (dollars per ton)	55.00	55.00	. . .
Efficiency measures			
Man-hours per ton	5.75	3.85	3.70
Capacity (millions of tons)	3	4	0.5
Yield to finished product (percent)	78	92	89

Source: Tables A-4, A-5, and A-6; data provided by steel engineers and producers.
a. Based on 90 percent utilization of production capacity.

is a collapse in demand for steel. How can this sector attract the required capital for growth at a time when the well-known, integrated steel companies are approaching bankruptcy? Obviously, investors must understand the prospects for minimill success in an industry awash with failures.

A 1984 study of minimill financing suggests that access to capital has been a problem for these firms.[6] In responses to a questionnaire, 37 percent of minimill respondents said that they thought that there is a serious shortage of affordable capital for new minimill projects. This would appear to be contradicted by the continuing growth in minimill

6. Zoltan J. Acs, "A Case Study of the Financing of Minimills in the U.S. Steel Industry: Implications for Public Policy," Working Paper (Manhattan College, School of Business Research Institute, November 1984).

capacity, but perhaps it is a common reaction of business executives to conditions in the capital market. Capital is always scarce.

Another study finds that the systematic risk of all steel investments in the 1960s and 1970s has been declining.[7] Even the riskiness of investment in the equities of integrated companies was less than the average for all risky assets by the late 1970s. In the 1970s both minimills and integrated firms were considered to be less risky than the average for all risky assets. This finding is puzzling, given the extreme difficulties in which the integrated firms find themselves, but perhaps it reflects the market's mistaken belief that trade protection would allow steel firms to earn an assured return on their existing assets, if not a very satisfactory return on new investment. The minimill companies have been expanding; hence, their rating must reflect the market's assessment of the risk of investing in new plants.

Among the major sources of new capital for minimills has been direct foreign investment, as the following list of companies and their foreign ownership interest shows:[8]

Company	Foreign ownership interest
Atlantic Steel Corp.	Ivaco (Canada)
Auburn Steel Co.	AC&Co.; Kyoei Steel (Japan)
Bayou Steel Corp.	Voest-Alpine (Austria); Getraco, Fimetal (France)
Copperweld Steel Co.	Sté Imétal (France) (66 percent)
Georgetown Steel Corp.	Established by Korf Industries (Germany)
Hawaiian Western Steel	Cominco (Canada)
Hurricane Industries	Otto Wolff America (Germany)
Laclede Steel	Ivaco (Canada) (50 percent)
New Jersey Steel Corp.	Von Roll (Switzerland)
Phoenix Steel Corp.	Creusot-Loire (51 percent) (France)
Raritan River Steel Co.	Co-Steel International (Canada)
Tamco	Tokyo Steel Manufacturing Co.; Mitsui & Co. (Japan)

Two of the companies—Phoenix Steel Corporation and Copperweld Steel Company—are not minimills in the strict sense, but neither are they integrated producers. Copperweld Steel specializes in alloy steels and Phoenix Steel produces plates with a configuration of electric

7. C. Y. Baldwin, J. J. Tribendis, and J. P. Clark, "The Evolution of Market Risk in the U.S. Steel Industry and Implications for Required Rates of Return," *Journal of Industrial Economics*, vol. 33 (September 1984), pp. 73–98.

8. Richard Serjeantson, Raymond Cordero, and Henry Cooke, eds., *Iron and Steel Works of the World*, 8th ed. (Surrey, England: Metal Bulletin Books, 1983).

furnaces and rolling mills that is not characteristic of minimills. Obviously, Japanese, French, Austrian, Swiss, and Canadian investors have been quite important in funding this expansion of the minimills. As long as this sector can continue to draw on diverse sources of international capital, it should have little difficulty in expanding to occupy the new product niches opened up by technological change.

Summary

The constant pressure for technological advances in minimill operations has produced substantial dividends. Technological change has been sufficiently rapid that the average costs of production in new mills have continued to decline despite high U.S. plant construction costs. This is in sharp contrast to the situation in the integrated sector, where new plants have become prohibitively expensive because they do not offer sufficient improvements in embedded technology to offset high capital costs.

Minimill technology has advanced to the point where minimills are prepared to challenge integrated firms in the production of sheet steels. Even with no major technological breakthroughs, small plants with a capacity of no more than 750,000 tons of sheet steel a year appear to offer lower costs than most of the integrated facilities. These new minimill operations may not be able to compete with integrated firms in markets where the metallurgical quality of the product is critical, but in lower-grade sheet products the invasion by the minimills appears imminent.

When mini hot-strip mills and the new technologies for casting thin slab have been implemented on a commercial scale, the minimill invasion of the sheet markets will become a matter of major concern to the large integrated firms. Only the most efficient integrated plants appear to be capable of surviving this attack.

V

Future Scrap and Electricity Supplies

Technological changes have been principally responsible for the twenty years of spectacular growth in the minimill sector of the steel industry. And those changes seem likely to continue to favor the small-scale producers. But will the material costs of these producers stay within manageable bounds? The answer lies in the availability and the price of steel scrap and, to a much smaller degree, of electricity.

Scrap and Its Substitutes

The success of the minimill in the United States depends very much on the continued flow of a supply of scrap or the development of substitutes such as directly reduced iron ore. By estimating how the determinants of scrap supply are likely to vary, it is possible to predict the limits on growth in electric furnace production in the next ten to fifteen years.

Scrap Supplies

The scrap market is highly competitive, with numerous buyers and sellers in each geographic market. Iron and steel scrap is generated by thousands of establishments. Steel fabricators who stamp, forge, extrude, and otherwise produce finished products lose large amounts of steel in trimmings, turnings, and rejects that are returned for recycling. Similarly, rejects and trimmings from the production processes of iron and steel foundries generate large quantities of scrap. Finally, there are thousands of scrap yards that collect and process "obsolete" scrap from discarded durable goods, store fronts, rails, bridges, and other structures. Obsolete scrap and the "prompt industrial" scrap generated by steel fabricators provide virtually all of the "purchased" scrap in the country.

In addition to purchased scrap, steel mills generate scrap in their

71

Table 5-1. *U.S. Scrap Supply, 1960–84*
Millions of tons

Year	Domestic consumption			Net scrap exports	Total U.S. scrap supply[c]
	Purchased scrap[a]	Home scrap	Total scrap[b]		
1960	26.1	39.6	66.5	7.2	73.4
1961	25.3	38.5	64.3	9.4	74.0
1962	25.3	40.6	66.2	4.9	71.3
1963	29.4	44.7	74.6	6.1	81.7
1964	31.8	52.3	84.6	7.6	92.5
1965	35.8	55.2	90.4	6.0	96.6
1966	36.7	55.5	91.6	5.5	97.8
1967	32.7	52.3	85.4	7.4	93.0
1968	33.6	53.5	87.0	6.4	93.4
1969	36.9	56.3	94.8	9.0	103.4
1970	34.1	52.6	85.6	10.6	95.8
1971	34.0	49.2	82.6	6.4	88.1
1972	41.7	51.2	93.4	7.4	98.0
1973	44.7	57.8	103.6	11.1	112.5
1974	51.3	55.3	105.5	8.8	109.4
1975	36.8	46.0	82.3	9.3	92.3
1976	41.4	50.0	89.9	7.7	97.7
1977	41.9	49.5	92.2	5.6	98.5
1978	46.1	52.1	99.2	8.3	108.1
1979	47.0	52.2	98.9	10.4	110.1
1980	41.0	42.2	83.7	10.8	95.1
1981	42.0	43.3	85.1	6.0	91.7
1982	28.0	27.1	56.4	6.5	63.3
1983	34.2	27.2	61.8	7.1	72.4
1984	36.1	29.3	65.7	9.3	74.0

Sources: U.S. Bureau of Mines, *Minerals Yearbook, 1984*, vol. 1: *Metals and Minerals* (Government Printing Office, 1985), pp. 548, 553, and earlier issues. Tons in all tables are net tons unless otherwise noted.
a. Receipts less shipments.
b. Includes inventory reductions.
c. Total scrap plus net exports less inventory adjustment.

production processes. This is often called "home" scrap. How much home scrap the steel plant produces depends on the volume of its raw steel output and the percentage that is lost in the fabrication process. Increased use of continuous casting and improved rolling practices have reduced yield losses in recent years. In the 1960s home scrap constituted about 60 percent of domestic consumption of scrap (table 5-1), but in the 1980s less than 50 percent. As continuous casting grows, the share of home scrap will fall even farther.

There is an extremely large network of scrap processors and dealers

Table 5-2. *U.S. Consumption of Scrap and Pig Iron, 1984*
Millions of tons

Furnace type	Scrap	Pig iron
Blast	2.7[a]	. . .
Basic oxygen	16.4	45.6
Open hearth	3.7	5.7
Electric	33.4	0.4
Cupola[b]	8.5	0.5
Other[c]	0.9	1.1
Total	65.7	53.2

Source: Bureau of Mines, *Metals and Minerals,* pp. 546–49.
a. Includes pig iron.
b. Furnaces used in foundries.
c. Includes direct castings.

in the United States. These enterprises collect, separate, and bundle scrap in various quality grades for delivery to steel plants and foundries. Many have facilities for processing obsolete automobiles and other durable goods, dismantling and crushing the discards, and separating the ferrous fraction through magnetic processes. The quality of scrap gathered in this fashion is often too poor for higher grades of steel or castings since trace elements in the steel or attached to the steel scrap are difficult to remove. Much of the output of minimills, however, does not require close metallurgical tolerances. In the future, as the minimills' production expands, both they and scrap dealers will increase their spending to sort scrap for higher-quality uses.

Demand for Scrap

Steel scrap is used by domestic and foreign foundries and steel mills. The number of buyers is extremely large and growing, as minimills become more numerous. Iron and steel foundries used only 13.9 million of the 65.7 million tons of scrap consumed in the United States in 1984. The rest was used by manufacturers of raw steel or pig iron.[1] As table 5-2 indicates, most scrap is consumed in electric arc steel furnaces.

Pig iron and directly reduced iron ore are substitutes for scrap, but not all facilities can use them interchangeably. Electric furnaces use virtually 100 percent scrap, though they can accommodate all three inputs. Other steel-making furnaces are more limited in the charges they

1. U.S. Bureau of Mines, *Minerals Yearbook, 1984,* vol. 1: *Metals and Minerals* (Government Printing Office, 1985), p. 549.

Table 5-3. *U.S. Raw Steel Production, by Furnace Type, 1961–85*
Millions of tons

Year	Open hearth	Basic oxygen	Electric	Total
1961	84.5	4.0	8.7	98.0[a]
1962	83.0	5.5	9.0	98.3[a]
1963	88.0	8.5	10.9	109.3[a]
1964	98.1	15.4	12.7	127.1[a]
1965	94.2	22.9	13.8	131.5[a]
1966	85.0	33.9	14.9	134.1
1967	70.7	41.4	15.1	127.2
1968	65.8	48.8	16.8	131.5
1969	60.9	60.2	20.1	141.3
1970	48.0	63.3	20.2	131.5
1971	35.6	63.9	20.9	120.4
1972	34.9	74.6	23.7	133.2
1973	39.8	83.2	27.8	150.8
1974	35.5	81.6	28.7	145.7
1975	22.2	71.8	22.7	116.6
1976	23.5	79.9	24.6	128.0
1977	20.0	77.4	27.9	125.3
1978	21.3	83.5	32.2	137.0
1979	19.2	83.3	33.9	136.3
1980	13.0	67.6	31.2	111.8
1981	13.5	73.2	34.1	120.8
1982	6.1	45.3	23.2	74.6
1983	6.0	52.0	26.6	84.6
1984	8.3	52.8	31.4	92.5
1985	6.4	51.9	29.9	88.3

Source: American Iron and Steel Institute, *Annual Statistical Report, 1985* (Washington, D.C.: AISI, 1986), table 27. Totals may not add because of rounding.
a. Includes some Bessemer production.

can handle. For instance, open hearth furnaces can utilize as much as 50 percent scrap, but basic oxygen furnaces generally do not charge more than one-third scrap unless the scrap is preheated.

The demand for scrap depends very much on the strength of the domestic steel market and electric furnaces' share of total production. Exports also constitute an important source of demand, though the high value of the dollar has diminished their role in recent years. In 1981 and 1982, exports were 40 percent below their 1979–80 peak (see table 5-1).

Electric furnaces have continued to increase their share of raw steel production, accounting for 20 percent in the early 1970s and more than 30 percent in 1985 (table 5-3). About one-third of this capacity is at integrated plants, where electric furnaces frequently have been installed to extend the useful economic life of aging facilities. Most of the increase in electric furnace capacity, however, is a reflection of the expansion of minimills.

Prices and Markets

With electric furnaces increasing their share of U.S. steel output, are scrap prices likely to be bid up sharply, and further expansion of minimills thereby made unprofitable? If recent price trends are any indication, this does not appear to be an immediate concern.

In the 1960s, nominal scrap prices were remarkably stable, and real (deflated) scrap prices actually fell (table 5-4). In 1973–74, a speculative boom gripped steel consumers, and scrap prices soared as a result. The real price of scrap rose by 150 percent in two years but fell back to early 1960 levels by 1977, as steel demand failed to recover. With U.S. steel output rising once again in 1978–79, the price of scrap rose once more, but it did not reach its 1974 real level. With the sharp decline in steel consumption in the 1980s, scrap prices fell back to the real levels of the 1960s.

Given the low value of steel scrap and the substantial costs of transporting it, the market for scrap cannot be considered national in scope. Prices differ between various regions of the country, and the decisions that minimills make about where to build their plants may be affected by these differences. For instance, Raritan River may have chosen its New Jersey location because of its plant's proximity to high-quality fabricated (prompt industrial) scrap in the Northeast.

The differences in scrap prices can be seen in the movements in the annual average price of number 1 heavy melting scrap in Chicago, Philadelphia, and Pittsburgh over more than two decades (table 5-5). Prices have differed by 10 or 15 percent across these three markets. For example, in 1981, the price of heavy melting scrap in Pittsburgh was nearly 15 percent above the average price in Philadelphia and 10 percent above the price in Chicago. Over time, however, the three price series are highly correlated, with zero-order correlation coefficients of 0.99 between each pair of price series.

A Model of the Steel Scrap Market

Given the numerous buyers and sellers of scrap and the gradable, homogeneous nature of the commodity, it is reasonable to assume that

Table 5-4. *Iron and Steel Scrap Prices, 1960–84*

Year	Heavy melting scrap (dollars per gross ton)	BLS iron and steel scrap price index	
		Index (1967 = 100)	Deflated index[a]
1960	32.95	110.2	126.8
1961	36.28	116.8	133.2
1962	28.24	95.2	106.6
1963	27.11	91.7	101.1
1964	33.67	109.4	118.8
1965	34.35	112.6	119.7
1966	30.93	106.6	109.8
1967	27.62	100.0	100.0
1968	25.85	93.0	89.1
1969	30.83	110.8	100.9
1970	41.06	138.8	120.0
1971	34.09	114.5	94.3
1972	36.92	121.8	96.3
1973	57.86	188.0	140.6
1974	108.49	353.2	242.6
1975	71.85	245.6	154.3
1976	77.74	259.0	154.7
1977	63.15	231.2	130.5
1978	76.34	264.6	139.1
1979	97.92	341.9	165.4
1980	91.34	328.1	145.4
1981	91.50	327.4	132.6
1982	63.13	232.9	88.0
1983	73.46	250.1	91.8
1984	86.06	288.1	102.0

Sources: David J. Joseph Co., "The Relationships between Scrap Prices, Steel Production, Purchased Scrap Receipts, and Scrap Exports, 1954–1983" (Cincinnati: The Company, 1984); prices of number 1 heavy melting scrap are from U.S. Bureau of Labor Statistics (BLS), *Producer Price Index*.

a. Deflated by GNP implicit price deflator (1967 = 100).

the market for iron and steel scrap is competitive. Markets clear after short lags, and speculative buying is difficult because of the large storage costs for the material.

The supply of scrap includes both purchased and home scrap. Home scrap is obviously a substitute for purchased scrap, and steel mills or foundries may sell it if excesses develop. Either product can satisfy the demand for scrap in steel furnaces or foundries.

The balance in the demand for and the supply of steel scrap at a given price will change as steel-making yields, the mix of metallic charges in

Table 5-5. *Average Price of Number 1 Heavy Melting Scrap in Three Cities, 1961–84*
Dollars per ton

Year	Chicago	Philadelphia	Pittsburgh
1961	35.19	38.69	35.22
1962	27.65	28.09	29.28
1963	27.96	25.79	26.93
1964	34.43	31.31	34.75
1965	34.14	33.60	35.10
1966	31.60	29.67	30.72
1967	27.96	27.98	26.95
1968	25.58	26.69	27.21
1969	29.21	30.82	31.61
1970	41.69	39.99	42.18
1971	33.49	33.19	36.71
1972	35.94	35.54	38.42
1973	57.28	58.11	57.61
1974	112.68	106.20	104.61
1975	71.36	74.20	72.01
1976	77.46	77.79	78.90
1977	60.21	65.53	66.42
1978	73.98	78.95	78.48
1979	96.69	98.61	100.77
1980	86.70	96.49	95.48
1981	91.76	87.67	100.57
1982	57.78	66.94	66.47
1983	72.42	69.31	76.99
1984	83.12	87.92	92.71

Source: American Metal Market, *Metal Statistics, 1985* (New York: Fairchild Publications, 1985), pp. 180–81; and *1980* issue, pp. 214–15.

furnaces, and the availability of obsolete scrap vary. If the mix of scrap and hot metal (pig iron) is not easily changed in a given furnace, and if yields in steel making and fabrication are not sensitive to the relative price of scrap, equilibrium in the scrap market depends heavily on changes in the rate of retrieval of obsolete scrap. And unless supplies of obsolete scrap are very elastic with respect to price, changes in any of the technological parameters could have a major impact on scrap prices.

A simple simulation is useful in demonstrating the importance of imports and the shift from open hearth to basic oxygen furnaces. This simulation, described in appendix B, solves for the share of basic oxygen and electric furnace production that equates the supply of and demand for scrap in a steady state in which steel consumption is not growing. The supply of scrap is a function of assumed yield ratios and recycling

Table 5-6. *The Sensitivity of Basic Oxygen and Electric Furnace Output Shares to Changes in Steel Imports and Open Hearth Production*[a]
Percent

Assumptions		Calculated shares	
Imports as a share of raw steel production[b]	*Open hearth share*	*Basic oxygen share*	*Electric furnace share*[c]
10	10	65	25
15	10	62	28
20	10	59	31
25	10	56	34
30	10	53	37
35	10	50	40
25	20	49	31
25	15	51	33
25	10	56	34
25	5	59	36
25	0	63	37

Source: Appendix B, equation 4. Assumed shares of imports and of open hearth production are used to calculate the shares of basic oxygen and electric furnace production.
a. Scrap exports held at 10 percent of U.S. supply and foundry output at 15 percent of raw steel production.
b. Steel exports are ignored.
c. Represents the maximum share for electric furnaces.

rates in steel production, steel fabrication, and final use. Imports of steel are also important because they add to the supply of scrap without requiring scrap for the original steel production. The demand for scrap depends upon the mix of open hearth, basic oxygen, and electric furnaces. As scrap supply increases, the share of electric furnaces in equilibrium must increase to absorb the scrap since electric furnaces consume a larger relative share of scrap than do basic oxygen furnaces. Table 5-6 demonstrates the changes in the share of steel produced in electric and basic oxygen furnaces as open hearth production and steel imports vary.

Clearly, the scrap supply, as reflected in potential electric furnace production, is more sensitive to changes in import penetration than to the limited range of changes that are still possible in open hearth production. At approximately the same conditions that applied in 1985, electric furnaces could account for 35 percent of U.S. raw steel production without requiring an increase in scrap prices to generate a larger supply of obsolete scrap. In a steady state, an increase in imports to 35 percent of U.S. raw steel production would increase the amount of scrap sufficiently to allow electric furnaces to increase their output to 40 percent of U.S. production. On the other hand, if open hearths disappeared completely, electric furnaces could only increase their output to 37 percent of raw steel output, all other things being equal. This obviously means that basic oxygen furnaces would increase their market share slightly if imports remained at 25 percent (assuming that the real price of scrap does not change).

It is often alleged that an increase in steel-making yields would reduce scrap supplies and thus reduce the market share for electric furnaces. But any increase in yield from raw steel to finished product would itself reduce the total demand for scrap because less raw steel would be needed to satisfy the demand for finished steel products. That explains why improvements in yields occasioned by continuous casting have not led to a rise in scrap prices.

Similarly, increases in fabrication yields should have only a modest effect on the equilibrium price for scrap. On the one hand, the increasing yields would reduce the scrap supply, but, on the other hand, they would reduce the overall demand for steel.

Finally, if steel consumption began to grow and the return rate for obsolete scrap stayed steady, then the supply of obsolete scrap as a share of current output would decline. This would place upward pressure on scrap prices. Obviously, if the scrap supply is to increase, the recycling rate must increase. The crucial question is how sensitive the price of scrap is to the recycling rate. That question is pertinent whether consumption is growing or not.

Estimates of Price Sensitivity of Scrap Supply

The supply of home scrap and prompt industrial scrap is likely to be directly proportional to activity in steel-using industries and not dependent on the relative price of scrap. As long as scrap prices are positive, it will pay steel-using industries to sell the scrap to processors. Some users might choose to manage scrap sales in a speculative manner, but the high cost of storage would seem to argue strongly against such behavior. Similarly, home scrap generated within steel plants is generally returned to the furnace with little delay. Thus supplies of prompt industrial and home scrap can be assumed to vary directly with activity

Figure 5-1. *Shifts in the Market for Scrap over the Business Cycle*

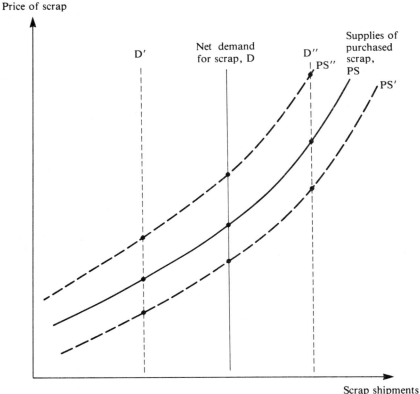

Price of scrap

Scrap shipments

in the steel and steel-fabricating industries and to be insensitive to scrap prices.

Supplies of obsolete scrap, on the other hand, are likely to be sensitive to scrap prices. Scrap yards can hold inventories of processed or unprocessed scrap, including old automobile hulks, appliances, or other steel materials. As scrap prices rise, the collection and processing of scrap into usable bundles for steel mills and foundries can be expected to rise also.

Figure 5-1 illustrates how supply and demand for scrap interact over time. The total net demand for scrap (*D*) is assumed to be a function of domestic steel output, world steel output, and the value of the dollar. This demand function moves in the range of *D'* to *D''* over the business cycle. Because the supply of home scrap is not significantly related to the price of scrap, only the supply function for purchased scrap (*PS*) is shown in figure 5-1. It shifts with steel output and consumption, moving in the range of *PS'* and *PS''* over the business cycle. (The equations for estimating these relationships are developed in appendix C.)

To forecast the availability and supply of scrap, it is important to have an estimate of the price sensitivity of the supply of purchased (or obsolete) scrap. A simple four-equation model of the U.S. scrap market was estimated using 1961–83 data (the estimates are reproduced in table C-1).

Forecasting Scrap Supply

The estimates of the relationship of purchased scrap supplies to the price of scrap suggest that at a given price, minimills can expect to find supplies of purchased scrap increasing by about 1 percent per year. Supplies could thus be expected to expand modestly throughout the 1980s without causing prices to rise above their late 1970s levels. While supplies of home scrap will continue to decline at integrated plants with the trend toward continuous casting, so will raw steel production. Integrated firms cannot increase their output of finished steel much above their 1978–79 rate for the rest of the 1980s without substantial trade protection or a highly inflationary boom. Their raw steel output is thus unlikely to return to 1978–79 levels in the 1980s.

Under reasonable assumptions about steel consumption, steel imports, scrap exports, foundry demand, and continuous casting, the prospects for scrap supply through the mid-1990s appear excellent. Electric furnace capacity should be able to increase from 30 percent, its mid-1980s share of production, to about 40 percent without any major increase in real scrap prices over their level in the late 1970s. Table 5-7 gives the results of simulations based on the econometric estimates in appendix C. The first set of estimates, using fixed coefficients for scrap demand in each type of furnace, is more useful than the second because it allows the shares of raw steel produced to be varied across furnace types.

Both sets of simulations show that scrap prices are not likely to rise appreciably above real 1979 levels even with electric furnaces turning out as much as 40 percent of the finished steel produced in the United States. Unless steel consumption rises substantially above 110 million tons, imports decline to less than 22 percent of U.S. steel consumption, or scrap exports rise above 12 million tons, domestic scrap prices are

Table 5-7. *Simulations of Scrap Consumption and Real Scrap Prices for 1995*[a]
Quantities in millions of tons

	Outcome, by simulation number							
Variable	*1*	*2*	*3*	*4*	*5*	*6*	*7*	*8*
Estimated values								
Assuming $DD = 1.6 \times FOUN + 1.05 \times ELEC + 0.27 \times [RAWSTEEL - ELEC]$								
Real scrap price (P)[b]	109	118	109	121	82	89	136	147
Scrap consumption								
(DD)	72.04	75.42	76.82	80.89	65.62	68.55	78.41	82.24
Assuming $DD = 6.25 - 1.08 \times \dfrac{P}{PIGP} + 0.456 \times RAWSTEEL + 1.60 \times FOUN$								
Real scrap price(P)[b]	85	92	82	90	85	90	87	93
Scrap consumption								
(DD)	66.61	69.20	70.38	73.51	66.98	69.58	66.28	68.88
Assumed values								
Steel consumption								
$(FABR)$	100	100	110	110	100	100	100	100
Steel imports (I)	27	22	30	24	27	22	27	22
Export demand for								
steel (ED)	2	2	2	2	2	2	2	2
Raw steel production								
$(RAWSTEEL)$	87.5	93.3	95.7	102.7	88.3	94.1	86.8	92.6
Electric furnace share								
$(ELEC)$	0.40	0.40	0.40	0.40	0.30	0.30	0.50	0.50
Steel produced by								
continuous casting								
$(CONCAST)$	66.5	70.9	72.7	78.1	64.6	67.7	69.4	74.1
Foundry shipments								
$(FOUN)$	13.2	13.2	13.2	13.2	13.2	13.2	13.2	13.2
Net scrap exports (FD)	12	12	12	12	12	12	12	12
Real price of pig iron								
$(PIGP)$[c]	56.46	56.46	56.46	56.46	56.46	56.46	56.46	56.46

a. Estimated from equations developed in appendixes B and C.
b. BLS index deflated by GNP deflator; $1979 = 100$.
c. Deflated by GNP deflator; $1979 = 100$.

not likely to rise more than 21 percent above their 1979 level. Since minimills were expanding in 1979, it seems likely that with similar real prices of scrap in the 1990s they would be able to prosper. This assumes, of course, that the real cost of producing hot metal in basic oxygen furnaces does not fall. Given the high price of domestic iron ore, the declining coking capacity in the United States, and the lack of investment in giant new blast furnaces, it seems likely that hot metal costs will not decline appreciably in real terms over the next decade.

Integrated plants, of course, might increase their use of scrap in the next decade. A number of technologies now under development, other than the usual preheating of scrap, would permit them to raise scrap charges in basic oxygen furnaces above the 26–27 percent in current practice. And the integrated firms might decide to rely more heavily on electric furnaces, as Bethlehem Steel and LTV have done at several aging plants. Also, some additional scrap could be charged directly into blast furnaces.

On the other hand, the integrated industry should close virtually all of its remaining open hearth capacity in the next few years. Since these furnaces consumed as much as 10.4 million tons of scrap per year as recently as 1979,[2] this will reduce scrap demand appreciably, offsetting any tendency of the integrated industry to rescue dying plants with electric furnaces.

The two remaining demands on U.S. scrap outside the minimill sector are from the export market and producers of steel and iron castings. Export demand for scrap obviously depends on the strength of the dollar and foreign steel production. Other countries are also increasing their electric furnace capacity. It seems unlikely, however, that the dollar will depreciate sufficiently to increase net scrap exports much above 12 million tons per year. Even at their peak, in 1973 and 1980, scrap exports were only 11.2 million tons.[3]

The castings industries depend very much on a number of durable goods industries. These industries will undoubtedly grow more rapidly than the basic steel industry, but not at a rate equal to the economic growth of the entire economy.

It therefore seems likely that at least 40 percent of all U.S. raw steel production could emanate from electric furnaces by 1995 with little real upward pressure on scrap prices. Minimills can and will compete in a wide range of products at the 1979 level of prices for scrap.

Substitutes for Scrap

With scrap prices depressed by the recessions of the early 1980s, interest in finding substitutes for scrap waned. Since there seems little

2. American Iron and Steel Institute, *Annual Statistical Report, 1982* (Washington, D.C.: AISI, 1983), table 45.
3. Bureau of Mines, *Minerals Yearbook, 1984*, p. 552.

likelihood of a repeat of 1974 scrap prices before the 1990s, there is little urgency in finding new sources of metallics for electric furnaces.

The principal alternative to scrap as a charge for electric furnaces is directly reduced iron (DRI). Though DRI was first produced commercially in 1957, it has never been widely used, principally because it requires large amounts of natural gas or gases produced from coal. Where market prices were charged for gaseous hydrocarbons, the energy cost of producing DRI was extremely high. As a result, the use of DRI has grown only in areas with abundant supplies of natural gas priced below world levels, such as Mexico, Venezuela, and the Middle East.

There is nearly 30 million tons of DRI capacity in the world, but as recently as 1982 world output was only 7.5 million tons. In the same year, total scrap consumption in noncommunist countries was nearly 170 million tons. Thus, DRI has barely begun to supplant scrap in electric furnaces. Moreover, its growth is almost entirely in scrap-poor countries. In the United States, Japan, Canada, and the European Community, DRI consumption is negligible and likely to remain very small. High energy costs, transportation problems, and the high price of DRI relative to scrap make it unattractive in developed countries. Only a major rise in scrap prices or a drop in gas prices would renew interest in DRI.

At current prices for scrap, electric furnaces using only scrap are clearly more economical to operate than those using a combination of scrap and directly reduced iron or than integrated plants that charge pig iron into basic oxygen furnaces. At the 1983 price of $73 per ton for scrap, the cost of producing liquid steel in an all-scrap electric furnace was $144, in an electric furnace using 65 percent DRI and 35 percent scrap $189, and in the basic oxygen furnace $167.[4] Those estimates, developed by the International Iron and Steel Institute, are somewhat below the estimate of raw steel costs in the low-cost U.S. minimill shown in appendix table A-2. When the price of scrap is assumed to be $100 per ton, the costs of the three types of production would be $174, $201, and $175, respectively. And if the price of scrap is put at $181 per ton—the real level reached in 1974—the costs of each type of production would be $263, $237, and $199, respectively. Obviously, scrap prices would have to be very high to encourage DRI-scrap combinations in electric furnaces in developed countries. Equally as important, the

4. International Iron and Steel Institute, *Scrap and the Steel Industry: Trends and Prospects* (Brussels: IISI, 1983), p. 9.30. Pig iron cost $141 per ton, DRI $127 per ton in 1983.

scrap-based electric furnace enjoys a clear advantage over integrated production when scrap prices are lower than $100.

The Outlook for Electricity

A minimill, by almost any definition, uses an electric furnace to melt scrap into liquid steel. While there is a possibility that technological change—perhaps spurred by a rise in the real costs of electricity—could make the electric furnace obsolete, such a change is not on the horizon. This means that the price of electricity will be of major concern to minimill operators for the foreseeable future.

The Role of Electricity Costs

Although electricity is the second largest identifiable cost (after scrap) in the operation of a minimill, it does not account for a large share of total costs. In 1985, electricity accounted for an estimated 12 percent of total costs and 14 percent of operating costs for efficient producers of wire rods (see table A-2). Some of this electricity is used to operate the rod mill, but 80 percent of it is consumed in the electric furnace. Technological progress is likely to reduce electricity consumption over time as minimills experiment with the use of different furnace and electrode designs. Nevertheless, a sharp increase in the cost of electricity relative to that of fossil fuels, particularly metallurgical coal, would reduce the future attractiveness of minimill investments. It is most likely, however, that electricity rates will fall as oil prices fall from their mid-1985 levels.

It is difficult to see how energy costs can be a serious competitive handicap to minimills over the long term. Producing steel by the blast furnace–basic oxygen furnace method consumes more than twice as much energy as electric furnace technology. Most of the energy in the integrated production method is contained in the coking coal consumed in coke ovens. If coking coal prices should escalate, a similar escalation could be expected in the steam coal that is increasingly the mainstay of the electric utility industry. If the costs of converting and distributing electricity should rise relative to coal prices, then the price of industrial electricity might also rise relative to coal prices, but this seems unlikely in the long run. Surely, productivity growth is likely to be greater in the

Table 5-8. *Producer Price Indexes for All Finished Products,*
Finished Energy Products, and Steel Mill Products, 1974–85
Base year 1967 = 100

Year	All finished products	Finished energy products[a]	Steel mill products
1974	147.5	215.2	170.0
1975	163.4	252.4	197.2
1976	170.6	282.3	209.8
1977	181.7	326.7	229.9
1978	195.9	347.7	254.4
1979	217.7	469.9	280.4
1980	247.0	701.3	302.7
1981	269.8	835.4	337.6
1982	280.7	822.9	349.7
1983	285.2	783.6	325.5
1984	291.2	750.3	366.0
1985	293.7	721.4	366.2

Sources: *Economic Report of the President, February 1986*, pp. 321, 323; AISI, *Annual Statistical Report, 1984*, table 5; U.S. Bureau of Labor Statistics, unpublished data.
a. Refined gasoline, natural gas, fuel oil no. 2, and finished lubricants.

generation and transmission of electricity than in the mining and transporting of coal. In the short run, however, electricity costs could rise
more rapidly than coal prices because of shortages in generating capacity
or changes in the policies of public utility commissions.

Recent Energy Price Trends

Since 1974, most energy prices have increased dramatically. By the
end of 1984, the producer price index for finished energy products was
7.5 times the 1967 level, while all finished products were only 2.9 times
their 1967 level. Since 1974, finished energy prices have risen nearly
twice as rapidly as the index for all finished products and the steel mill
products component of the index (see table 5-8). Not surprisingly, energy
costs for steel makers have risen at a similar rate, regardless of the
technique of production employed.

According to the U.S. Department of Energy, industrial electricity
rates have quadrupled since 1973 (table 5-9). The pattern of the increase
has been affected by regulation. Regulators succeeded in keeping electricity price increases below the average rate of energy price inflation in

Table 5-9. *Prices of Electricity, Coking Coal, and Natural Gas,*
1973–85
Dollars

Year	Industrial electricity (cost per kilowatt hour)	Coking coal (cost per ton)	Natural gas (cost per thousand cubic feet)
1973	0.0125	19.11	0.58
1974	0.0169	33.00	0.69
1975	0.0207	45.98	0.96
1976	0.0221	48.37	1.30
1977	0.0250	50.17	1.83
1978	0.0279	52.75	2.17
1979	0.0305	54.50	2.84
1980	0.0369	56.04	3.76
1981	0.0429	59.34	4.00
1982	0.0495	60.86	4.05
1983	0.0497	51.75	4.35
1984	0.0503	53.32	4.70
1985	0.0517	53.07	4.60

Sources: U.S. Department of Energy, Energy Information Agency, *Monthly Energy Review,* January 1986, p. 100; Peter F. Marcus, Karlis M. Kirsis, and Donald F. Barnett, *World Steel Dynamics: The Steel Strategist #11* (New York: Paine Webber, 1985), tables 24, 26.

the early to mid-1970s.[5] Since 1980, however, the price of industrial electricity has risen much more rapidly than other energy prices. In 1980–82 alone, industrial electricity rates increased by 34 percent while the average price of energy in the finished goods segment of the producer price index rose by only 17 percent. The rise appears to have moderated: between 1983 and 1985 industrial electricity rates increased by an average of 1.5 percent a year.

Table 5-9 shows that between 1973 and 1985 coal prices rose substantially less than electricity rates or natural gas prices; therefore one might expect energy costs to have risen more rapidly in minimills than in integrated plants. Since integrated plants use an average of about one ton of metallurgical coal per ton of finished product and substantial amounts of electricity, steam coal, and natural gas, their energy costs are the sum of all of these components. Since 1973, the increase in energy costs for integrated producers has been substantial (see table 5-10). In 1985, at actual operating rates, it appears that energy costs per ton in integrated production were 4.25 times their 1973 level in current dollars.

5. See Andrew S. Carron and Paul W. MacAvoy, *The Decline of Service in the Regulated Industries* (Washington, D.C.: American Enterprise Institute, 1981), p. 44.

Table 5-10. *Cost of Energy for Producing Steel in Integrated Plants and Minimills, 1973–85*
Dollars per ton of finished product

Year	Integrated plants[a]	Minimills[b]
1973	17.72	11.54
1974	22.38	15.06
1975	29.23	18.52
1976	33.20	20.28
1977	38.43	23.66
1978	45.12	26.23
1979	49.07	29.69
1980	56.89	36.21
1981	64.48	40.44
1982	77.06	44.35
1983	76.83	44.58
1984	76.92	45.17
1985	75.32[c]	45.08[c]

Sources: Same as table 5-9.
a. Cost for full mix of products, based on metallurgical coal costs plus two-thirds of other energy costs (to allow for coke-oven gas credits).
b. Cost for wire rods, assuming natural gas use declines loglinearly from 3.20 million British thermal units per ton in 1973 to 2.55 million in 1985; electricity use declines from 775 kilowatt hours per ton in 1973 to 645 in 1985.
c. Authors' projections.

In 1985 an efficient minimill producing wire rod consumed about 645 kilowatt hours of electricity per finished ton at the electric furnace and the rolling mill and 2,550 cubic feet of natural gas at various stages of production. The total energy cost per ton of wire rods at a minimill rose from $11.54 in 1973 to $45.08 in 1985. This rate of increase is less than the rise in integrated firms' energy costs because of substantial investments in energy efficiency in the minimills.

Future Supply of Electricity

Despite the obvious advantages that minimills enjoy in energy costs, there is some concern that the supply of electrical energy could fail to keep pace with demand in future years, causing a significant shortage that would require a subsequent sharp increase in rates or rationing of "nonessential" uses such as steel production. This concern is grounded in the difficulties that electrical utilities faced during a period of inflation, high interest rates, demand uncertainty, and hostile regulation of public utilities.

Table 5-11. *Various Forecasts of Growth in Demand for Electricity*

		Annual growth rate (percent)	
Forecaster and date of forecast	*Growth period*	*Gross national product*	*Electricity demand*
U.S. Department of Energy			
Energy Information Administration, April			
1983	1982–1990	3.2	3.8
Office of Policy Planning and Analysis,			
October 1983	1982–2000	2.8	2.8
June 1983	1982–2000	3.0	3.0
Wharton Econometrics, August 1983	1982–2000	2.9	2.8
Data Resources, Inc., Autumn 1983	1982–2000	2.8	2.4
U.S. Department of Commerce (Gustaferro)			
July 1983	1982–2000	2.8	2.3
North American Electric Reliability Council			
July 1983	1983–1992	. . .	2.8
Early 1985	1985–1994	. . .	2.2
Congressional Research Service			
August 1983	1982–2000	3.0	2.6
William Hogan			
April 1984	1983–2000	3.0[a]	2.3
		2.0[a]	1.4

Sources: *Need for New Powerplants,* Hearings before the Subcommittee on Energy Conservation and Power of the House Committee on Energy and Commerce, 98 Cong. 2 sess. (Government Printing Office, 1984), pp. 184, 322; William W. Hogan, "Patterns of Energy Use," in John C. Sawhill and Richard Cotton, eds., *Energy Conservation: Successes and Failures* (Brookings, 1986), pp. 42–44.

a. Rate of growth chosen, not a forecast.

Historically, utilities have tended to overestimate the need for future capacity. Because of rising real energy prices and sluggish economic growth since the early 1970s, most forecasts of the demand for electricity have proved exceedingly optimistic. Most recently, however, studies of demand growth have concluded that electricity sales will grow at about the same rate as the gross national product or even at a slower rate (table 5-11). The Congressional Research Service and Data Resources have forecast electricity demand growth for the rest of this century to be only 60–85 percent of GNP growth, given moderate growth in real prices. And Hogan asserts that even given *stable* prices, this figure will only be 70–80 percent.[6]

6. Congressional Research Service, *A Perspective on Electric Utility Capacity Planning,* prepared for the Subcommittee on Energy Conservation and Power of the House Committee on Energy and Commerce (Government Printing Office, 1983); Data Resources, Inc., *Energy Review,* quarterly issues; William W. Hogan, "Patterns of

Much of the debate concerning demand turns on the speed at which users adjust to price increases and the nature of new energy-using technology. If the adjustment is relatively slow and if many of the new energy-saving technologies are also electricity saving, electricity demand may grow more slowly than GNP for a long time. On the other hand, if the need to save energy encourages switching to electricity, as in the case of minimills, and if the adjustment to the two 1970s oil shocks is already relatively complete, demand for electricity could resume its pattern of growing more rapidly than GNP.

The prospective long-run growth rate of demand may be irrelevant to the availability and price of electricity. Utilities should be able to adjust their capital spending plans to account for any prospective demand pattern. If the long-run supply of electricity is highly price elastic, it might not matter whether demand grows at 2 percent or 4 percent a year. With a very elastic supply of coal, fuel costs should not contribute to sharp increases in costs. The location of new generating plants could be a source of increasing costs, and so could environmental controls biased against new plants. But the electric utility industry gives no evidence of conditions that would suggest rapidly increasing costs over the foreseeable range of output.

The principal hazard, then, is that utilities will not be able to or may not find it attractive to increase capacity. Whether the political and economic climate might impede investment in new plants in the next twenty or thirty years is difficult to predict. However, the estimates of planned additions to generating capacity through 1994 provide little basis for concern.

The North American Electric Reliability Council's estimates of generating capacity, or resources, through 1994 are based on the utilities' own planned additions, facilities under construction, existing capacity, and planned retirements. The utilities' forecasts of peak demand over that period have been going down steadily. In 1982, the average ten-year forecast of the growth rate in peak demand was 3.0 percent. In 1983, it fell to 2.8 percent, in 1984 to 2.5 percent, and in 1985 to 2.2 percent.[7]

Projections of the adequacy of electrical generating capacity must be based on regional analyses. The forecasts of resources for 1990 and 1994

Energy Use," in John C. Sawhill and Richard Cotton, eds., *Energy Conservation: Successes and Failures* (Brookings, 1986).

7. North American Electric Reliability Council, *Electric Power Supply and Demand, 1985–1994* (Princeton, N.J.: NERC, 1985), p. 9; *1984–1993*, p. 10; *1983–1992*, p. 8.

Table 5-12. *Projected Summer Peak in Electricity Supply and Demand, by Region, 1990 and 1994*

Projection period and region[a]	Planned resources (NERC) (thousands of megawatts)	Effective safe capacity (thousands of megawatts)[b]	Peak summer demand			
			NERC		CRS	
			Estimate (thousands of megawatts)	Percent of safe capacity	Estimate (thousands of megawatts)	Percent of safe capacity
1990 projections						
ECAR	98.9	79.1	75.5	95.4	78.0	98.6
ERCOT	55.6	44.5	45.4	102.0	44.3	99.6
MAAC	49.7	39.8	37.5	94.2	36.6	92.0
MAIN	49.5	39.6	38.0	95.6	37.7	95.2
MAPP-US	30.2	24.2	24.5	101.2	24.9	102.9
NPCC-US	57.2	45.8	42.5	92.8	38.3	83.6
SERC	144.2	115.4	110.8	96.0	106.5	92.3
SPP	66.7	53.4	53.8	100.7	54.5	102.1
WSCC-US	126.8	101.4	91.7	90.4	89.6	88.4
All regions	678.8	543.2	519.7	95.7	510.4	94.0
1994 projections						
ECAR	104.1	83.3	81.2	97.5	. . .	. . .
ERCOT	63.1	50.5	52.4	103.8	. . .	. . .
MAAC	50.2	40.2	39.3	97.8	. . .	. . .
MAIN	50.0	40.0	40.5	101.3	. . .	. . .
MAPP-US	30.6	24.5	26.5	108.2	. . .	. . .
NPCC-US	59.5	47.6	45.5	95.6	. . .	. . .
SERC	153.4	122.7	121.9	99.3	. . .	. . .
SPP	71.9	57.5	59.4	103.3	. . .	. . .
WSCC-US	134.0	107.2	100.1	93.4	. . .	. . .
All regions	716.8	573.5	566.8	98.8	. . .	. . .

Sources: North American Electric Reliability Council, *Electric Power Supply and Demand, 1985–1994* (Princeton, N.J.: NERC, 1985), pp. 9, 13; Congressional Research Service (CRS), *A Perspective on Electric Utility Capacity Planning,* prepared for the Subcommittee on Energy Conservation and Power of the House Committee on Energy and Commerce, 98 Cong. 1 sess. (GPO, 1983), p. 8.

a. The NERC regions are East Central Area Reliability Coordination Agreement, Electric Reliability Council of Texas, Mid-Atlantic Area Council, Mid-America Interpool Network, Mid-continent Area Power Pool, Northeast Power Coordinating Council, Southeastern Electric Reliability Council, Southwest Power Pool, and Western Systems Coordinating Council.

b. Figured as 80 percent of planned resources.

in the nine NERC regions (see table 5-12) include a reserve margin required to assure stable operation of the utility. This reserve is generally assumed to be 20 percent of capacity; the effective safe capacity therefore represents 80 percent of forecast resources.

The demand forecast from the Congressional Research Service (CRS) assumes a 2.6 percent growth annually in peak demand, somewhat above the NERC forecast (presumably because it was made two years earlier). Neither forecast provides much basis for concern through 1990. There are minor shortfalls in only three regions (two, using the CRS demand estimate)—the upper plains states (MAPP), the south central states

Figure 5-2. *Safe Capacities and Margins for Meeting Peak Demand for Electricity in the United States, 1954–92*

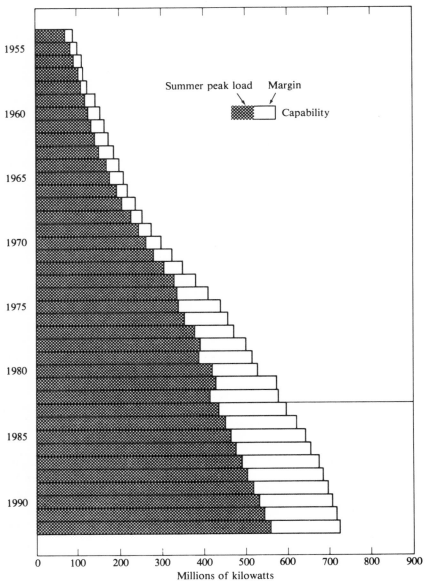

Millions of kilowatts

Source: Edison Electric Institute, *1983 Annual Electric Power Survey* (Washington, D.C.: EEI, 1984), p. 10; figures beyond 1982 are projected.

(SPP), and Texas (ERCOT). None of the shortfalls is very large, amounting to about 2 percent over safe capacity at most.

By 1994, a few more minor problems emerge, including a shortfall in the Illinois-Michigan-Missouri area (MAIN). But the fragility of ten-year forecasts and the tendency of utilities to overestimate demand make it premature to suggest that there might be price pressures or nonprice rationing of electricity in these areas.

The improvement in the average margin of capability for all electric utilities in the United States to meet peak demand is graphically demonstrated in figure 5-2. After declining in the late 1960s and early 1970s, the average rose sharply—in large part because of the reduction in demand growth caused by the first oil-price shock. Since 1975, reserve margins have remained between 30 and 40 percent of safe capacity, far above the average for the 1950s or 1960s. Moreover, the Edison Electric Institute, relying on NERC data, predicts that these margins will remain above 30 percent until the early 1990s.[8]

Price Forecasts

In an unregulated industry, forecasts of demand and capacity would be directly related to price forecasts. In a regulated industry with long-lived assets, however, prices can be restrained by regulators even when there is a growing imbalance between expansions of capacity and growth in demand. In the electric utility industry, nonprice rationing and voltage reductions can replace the price mechanism if demand begins to outrun supply at the regulated rates. In fact, during the 1960s the electric utility industry's capability margin declined steadily, leading to numerous power failures in the early 1970s. This reduction in capability margins has been attributed to the failure of public utility commissions to allow utility rates to rise with costs.[9] Since 1973, Carron and MacAvoy report, the commissions have been more favorably disposed to allowing rate increases in response to cost inflation. Moreover, the sharp decline in demand growth, caused by sharply higher energy prices since 1973–74, has allowed capability margins to rise once again.

In 1980–82, electric utility rates rose at an average annual rate of 14 percent. Industrial rates rose even more sharply—at an average rate of

8. Carron and MacAvoy, *Decline of Service,* pp. 41–42.
9. Department of Energy, Energy Information Administration, *Monthly Energy Review,* January 1986, p. 100.

16 percent. Much of this increase was in response to rising energy costs caused by the second oil-price shock. Since 1982, electricity prices have moderated substantially, and industrial rates have actually fallen. Unless demand begins to grow so rapidly that utilities begin to operate with reserve margins very near 20 percent once again, utility commissions will not be under pressure to increase rates to allow utilities to raise capital for expansion. Even if rates are not generally high enough to cover the costs of developing new power sources, they may not rise appreciably in the foreseeable future.

In a 1983 study assessing future demand, the Department of Energy first assumed an annual real growth rate of utility prices of 1.67 percent through 2000, but later revised this assumption downward to 0.5 percent.[10] The 1983 CRS report assumes a 2.0 percent real growth rate.[11] And a detailed analysis submitted to the Department of Energy predicts a growth rate of 0.3 to 1.4 percent per year, depending on the strength of the economy, inflation, regulatory policies, world oil prices, and utility investment strategies.[12] Obviously, forecasting energy prices is a hazardous specialty, but the consensus until recently has been that rates will rise more rapidly than the general price level, though the difference will not be large. The 1986 drop in oil prices may change this consensus.

The only disquieting aspect of the electricity market is the possibility that regulatory recalcitrance in response to strong growth could place substantial strains on generating capacity in the Southwest. This would make it difficult to assure future sources of power for large new industries in the region, unless utilities can import electricity from adjacent regions. The rapidly growing Southwest should be a major market for minimill producers, given their specialization in construction products.

Summary

At this point in the history of minimills, there is no reason why either scrap supply or electricity rates or supply should pose a barrier to their

10. *Need for New Powerplants,* Hearings before the Subcommittee on Energy Conservation and Power of the House Committee on Energy and Commerce, 98 Cong. 2 sess. (GPO, 1984), pp. 22–33, 34–68.

11. CRS, *Perspective on Electric Utility Capacity,* p. 7.

12. ICF, Inc., *An Analysis of the Electric Utility Industry in the U.S.: An Overview of Selected Analyses Assessing the Financial Health of the Utility Industry,* prepared for the Department of Energy, Office of Policy, Planning and Analysis, Electric Utility Division (The Division, 1983), p. 18.

expansion. Any increase in energy prices generally will discomfit integrated producers more than minimills. If regulatory policies constrain electricity growth, new minimills could find it difficult to obtain assured contracts for power, but none of the forecasts from the early 1980s has predicted supply shortages by the early 1990s. Mildly rising real electricity rates will increase minimills' energy costs, but they will not provide a crippling blow to expansion. These firms' electricity costs are now only about 14 percent of operating costs. Energy-saving technology will undoubtedly offset any rise in real electricity rates, leaving this share no higher in 1990 than it is today.

Though the use of purchased scrap can pose severe quality problems in steel making, most of the problems can be mitigated by careful sorting and screening or by corrective measures at the furnace or in the ladle. Enough scrap exists in the United States to allow minimills to double their 1985 capacity by the late 1990s, a rate of expansion that is quite optimistic given the rate of growth of demand for steel mill products.

VI

Industrial Policy: The Lessons from Steel

The recent problems of the U.S. steel industry have been felt most intensively in the integrated sector, where capacity for making steel has been reduced sharply. The cuts are even deeper than they appear to be, for many plants will never resume capacity operation. The problems of the integrated sector have, of course, been exacerbated by the growth and competitiveness of the minimills, which are still adding capacity and expanding their product range. However, the minimills too have problems. They have tended to overcrowd in the product markets they find most comfortable: reinforcing bars and light shapes. The minimills have found it easy to enter the low-product-quality range, and now they must realize the corollary—it is also easy (but not painless) to exit. A shakeout in the minimill sector is under way and is likely to continue, unless some companies move to other products.

The Economic Climate for Minimills

In the coming years, growth in the steel market will be slow, both worldwide and in the United States. This will be true in the product specialties of both minimills and integrated firms. Thus, minimills will grow largely at the expense of integrated producers.

Projections of steel consumption are based on an analysis of the rate of growth of the demand for steel relative to gross national product (in 1972 dollars). The ratio of steel consumption to GNP has been falling continually in the United States since the 1950s.[1] In 1955–59 the United States consumed 108.5 tons of steel per million dollars of real GNP. By 1970–74, consumption had declined to 92.9 tons, even though there had

1. American Iron and Steel Institute, *Annual Statistical Report, 1984* (Washington, D.C.: AISI, 1985), table 1A and earlier issues; *Economic Report of the President, February 1985*, p. 234.

been no upward movement in the real price of energy from most sources. And by 1980–84, steel consumption had fallen to 60.2 tons per million dollars of GNP, as higher energy prices forced producers and consumers to shift away from heavy steel products and the high value of the dollar increased imports of durable goods made from steel.

The rate of decline in the ratio of steel consumption to real GNP was 1.0 percent between 1955–59 and 1970–74 and 4.3 percent between 1970–74 and 1980–84. With the fall in real energy prices in the mid-1980s, the rate of decline should moderate somewhat, particularly if the dollar remains weak. The low projections of steel consumption in 1990 and 2000 shown in table 6-1 assume that this rate will be 3.0 percent between 1985 and 2000, while the high projections posit a 2.3 percent rate of decline.[2]

International competition can be expected to become even more keen as foreign minimills expand. Minimills are likely to become increasingly important throughout the world; indeed, electric furnace capacity should expand worldwide from 24 percent capacity in the mid-1980s to over 30 percent by the mid-1990s. As governments and private investors in developing countries recognize that most markets are not large enough to justify an integrated mill, minimills will be allowed to grow with the market. While the scrap they rely on may often have to be imported, the cost penalty will be less than the cost of continuing to keep huge, overbuilt integrated plants. Unfortunately, direct reduction of iron ore has never been successfully integrated into minimill processes at a reasonable operating cost; some new technology using iron ore will have to replace this technique unless its costs can be reduced substantially. Recent energy price declines, however, open the possibility of a major decrease in the cost of directly reduced iron ore.

The growth of international competition, compounded by slow market growth, will pose a major continuing threat to the U.S. market, even as the U.S. dollar recedes from its 1985 level. With the dollar's 1985–86 decline, import shares in the U.S. market should stabilize at 25 percent and perhaps even decline, but they will resume their slow growth to 30 percent of the U.S. market by 2000. The growth in import share will be restrained by the increased competitiveness of the U.S. industry, mini-

2. The forecast of steel shipments is based on the Wharton long-term forecast of GNP extrapolated to 2000 by the 1985–95 growth rate. This provides an estimate of GNP for 2000 of $2,560 billion (in 1972 dollars). The more optimistic scenario assumes a relatively weak dollar and, therefore, a strong steel fabrication sector.

Table 6-1. *Projections of Growth in the U.S. Carbon Steel Industry, Various Years, 1980–2000*

			1990		2000	
Item	*1980*[a]	*1985*	*Low*	*High*	*Low*	*High*
Steel consumption (millions of tons)[b]	101.7	96.4	94.0	98.0	95.0	105.0
Import share of U.S. market (percent)	17.8	25.2	25.0	25.0	30.0	30.0
Total imports (millions of tons)	18.1	24.3	23.5	24.5	28.5	31.5
Domestic shipments (millions of tons)	83.6	72.1	70.5	73.5	66.5	73.5
Exports (millions of tons)	2.8	0.9	1.0	1.0	1.0	1.0
Total net shipments (millions of tons)	86.4	73.0	71.5	74.5	67.5	74.5
Minimills						
Share of total net shipments (percent)	13.9	21.0	27.0	26.0	40.0	40.0
Total minimill shipments (millions of tons)	12.0	15.3	19.3	19.3	27.0	29.8
Finished product yield (percent)	0.85	0.87	0.88	0.88	0.92	0.92
Raw steel production (millions of tons)	14.0	17.6	21.9	21.9	29.3	32.4
Raw steel capacity (millions of tons)	16.0	22.0	25.0	25.0	33.5	37.0
Labor productivity (man-hours per ton)	4.0	2.8	2.2	2.2	1.5	1.5
Integrated and specialty firms						
Total net shipments (millions of tons)	74.4	57.7	52.2	55.2	40.5	44.7
Finished product yield (percent)	0.73	0.82	0.81	0.81	0.87	0.87
Raw steel production (millions of tons)	102.1	70.7	64.4	68.1	46.6	51.4
Raw steel capacity (millions of tons)	139.0	111.6	75.8	80.1	54.8	60.4
Labor productivity (man-hours per ton)	9.5	7.1	6.0	6.0	4.8	4.8
All steel manufacturers						
Raw steel production (millions of tons)	116.1	88.3	86.3	90.0	75.9	83.8
Raw steel capacity (millions of tons)	155.0	133.6	100.8	105.1	88.3	97.4
Capacity utilization (percent)	74.9	66.1	85.6	85.6	86.0	86.0
Labor productivity (man-hours per ton)	9.4	6.2	5.0	5.0	3.5	3.5
Number of employees (thousands)	429.5[c]	239.3[c]	188.2	196.1	124.3	137.2

Sources: Authors' projections; see text for explanation. Data for 1980 and 1985 from American Iron and Steel Institute, *Annual Statistical Report, 1985* (Washington, D.C.: AISI, 1986), tables 1A, 1B, and *1982* issue. Tons in all tables are net tons unless otherwise noted.

a. Five-year average from 1978–82.

b. Apparent steel consumption, which does not take into account changes in inventories. Includes domestic shipments and imports of semifinished steel, which are projected to rise to 5 million tons per year by 2000.

c. AISI total employment, adjusted for share of output represented in AISI, *Annual Statistical Report*, various years.

mills and integrated firms alike, if the dollar remains near mid-1986 levels.

Scrap shortages are not expected to be a barrier to the projected expansion of the minimill sector in the United States. The United States will continue to import steel, cars, and other products in sufficient quantities to supply scrap for both the domestic market and exports. The real price of scrap will rise, but not enough to offset the other advantages minimills have. As a result, minimills should account for approximately 40 percent of U.S. raw steel production by 2000. This is consistent with the projected potential of 40 percent of steel production in electric furnaces and the complete phaseout of the integrated firms' electric furnaces.

Other countries will have more severe scrap problems than the United States, particularly those selling steel and motor vehicles to the United States. In areas where steel consumption is not expected to grow, such as the European Community, scrap availability will approach that in the United States, encouraging major inroads by minimills.

Technological changes are likely to favor minimills since they will reduce the minimum economic scale necessary to produce steel efficiently and make it possible to produce higher-quality goods from scrap, while possibly also creating a scrap substitute of reasonable cost. These developments will favor the growth of minimills worldwide, although the advantage of minimills over their integrated competitors will be considerably less in other countries than in the United States because scrap supplies are less ample in other countries, labor costs lower (which reduces the importance of lower labor use in minimills), and iron ore cheaper. By the 1990s these technological changes may cause the distinction between integrated and nonintegrated steel mills to fade, as minimills and integrated producers move toward medium-scale plants (bigger than minimills, smaller than integrated plants), with new types of melting furnaces, all using a mix of iron ore and scrap.

The combined impact of all of these changes will be to reduce U.S. raw steel capacity from 134 million tons in 1985 to between 88 million and 97 million tons by the end of the century. This implies that capacity in the integrated sector will decline by nearly 50 percent in this fifteen-year period as minimills grow slowly and imports rise gradually in a market with essentially no growth. Even if steel consumption were to grow by as much as 1.7 percent per year, the necessary capacity at integrated plants would be no more than 70 million tons in 2000. Thus, even with these optimistic assumptions, the integrated firms would have to reduce their capacity by 40 million tons in the next fifteen years. The high estimate for consumption in 2000 shown in table 6-1 implies the closure of more than 50 million tons of the integrated firms' 1985 capacity.

Employment in the U.S. steel industry has declined by 60 percent since 1965, to 240,000 in 1985, and by 2000 will decline to between 124,000 and 137,000. Only by large improvements in productivity can U.S. producers keep 70 percent of the U.S. market, but these improvements obviously translate into substantially lower employment. Thus, even if output grows modestly, employment will continue to fall very rapidly.

Import Restrictions

Minimills and integrated producers alike are feeling the problem of increased import competition in the United States. This is a reflection of the slow growth in demand worldwide and the excess capacity it has generated, plus the strong rise in the U.S. dollar through 1985. Minimills are much better positioned, in terms of costs, to meet the foreign competition, although some may not have the financial resources to outlast their foreign rivals.

In January 1984, Bethlehem Steel and the United Steelworkers filed an escape clause petition (section 201 of the Trade Act of 1974), asking for a quota on imported steel of 15 percent of U.S. consumption. After determining that the industry was injured, the International Trade Commission recommended that tariffs and quotas be applied to some imported steel products. The president rejected formal quotas and tariffs, but instead agreed to negotiate "voluntary export" agreements with steel-exporting countries in October of the same year. These restrictions would reduce imports from 26.4 percent of U.S. consumption in 1984 to perhaps 21 percent in 1986. Though imposition of import restrictions may reduce the pressure on U.S. steel makers somewhat, it will not prevent the minimills from solving their problems of excess capacity by continuing to invade the integrated producers' markets.

Minimill Strategies and Market Penetration

The growth of the minimills, in a period of nearly catastrophic decline for the U.S. integrated companies, is due to a combination of factors such as low scrap prices; improvements in electric furnaces; efficient, small-scale plant design; low labor costs; and carefully chosen geographic markets for narrow ranges of products. Indeed, those minimills that have encountered difficulties have generally failed to choose a narrow product range or target their product to a narrow geographic region and have built plants that are simply too large to serve a small geographic area efficiently.

The successful companies, however, are growing and expanding their capacity. Nucor is introducing new technologies and invading product markets by building new plants. North Star, on the other hand, has been buying plants from minimill companies in financial distress. Relatively few plants have been scrapped, and many plants have been built in the past ten years. As minimills succeed in pioneering new technologies, we would expect them to continue to expand capacity even in a no-growth market.

Predicted Shares

In the fifteen-year period between 1985 and 2000, minimills will expand their share of the U.S. market under almost any condition other than extremely rapid growth in steel demand. They will be able to do this because integrated firms will continue to abandon facilities that compete with minimills, and because the minimills will move into new product lines, such as sheet, plate, and welded pipe. In addition, small nonintegrated steel companies may be established simply to produce slabs for rolling mills owned by the integrated companies or successor firms.

By continuing to exploit the efficiencies of small-scale electric furnace operation and constantly updating their plants to reflect the latest technology, the minimill sector will realize substantial growth. Current trends in technology support the forecasts contained in table 6-2. Clearly, the smaller companies will dominate production of small-diameter products by 1990, but they will also account for a substantial share of production of large structural shapes and seamless tubing. Their extension into the flat-rolled products, however, will depend on technological breakthroughs whose timing is difficult to predict. In addition, the supply of scrap may begin to become a constraint when about 40 percent of domestic shipments are from minimills. By 2000, however, there undoubtedly will be either major changes in the technology of using iron ore in electric furnaces or some modification of the basic oxygen converter to allow iron and scrap charging without the costly coke ovens

Table 6-2. *Estimated Minimill Share of U.S. Steel Shipments,
Various Years, 1980–2000*

Product	Minimill share				
	1980	1985	1990	1995	2000
	Percent				
Semifinished slabs and billets	10.0	20.0	30.0	40.0	50.0
Wire rods	45.0	80.0	90.0	95.0	100.0
Merchant bars	37.5	60.0	80.0	90.0	100.0
Light shapes	92.0	95.0	100.0	100.0	100.0
Reinforcing bars	75.0	95.0	100.0	100.0	100.0
Cold-finished bars	15.0	25.0	40.0	60.0	75.0
Structural shapes over 3 inches in cross section	30.0	45.0	60.0	70.0	80.0
Plate	. . .	. . .	7.5	20.0	35.0
Rails	. . .	. . .	30.0	65.0	100.0
Track material, wheels, and axles	. . .	15.0	20.0	25.0	35.0
Hot-rolled sheet	. . .	. . .	2.5	15.0	20.0
Cold-rolled sheet	. . .	. . .	. . .	5.0	10.0
Galvanized sheet	. . .	. . .	. . .	5.0	10.0
Pipe and tube					
Seamless	. . .	15.0	30.0	50.0	60.0
Welded	. . .	. . .	5.0	10.0	15.0
All products	13.8	21.0	26.5[a]	32.4	40.0[a]
	Millions of tons				
Addendum					
Total net shipments	86.4	73.0	73.0[a]	72.0	71.0[a]
Total minimill capacity	16.0	22.0	25.0	28.9	35.2

Source: Authors' estimates.
a. Midway between the high and low estimates in table 6-1.

and blast furnaces. Thus, to predict more than fifteen years ahead would
be both risky and foolish.

Some consolidation of minimill producers is to be expected, reducing
the number of current independent producers and giving the remaining
producers a more diversified market with no reduction in plant specialization. Indeed, the recent emergence of micromills (less than 100,000
tons) points to greater specialization within a very narrow product range
and a very narrow size range (for example, merchant bars 0.50 to 0.75
inch in diameter) to serve a small geographic area.

Given the current excess capacity in standard bars, shapes, and rods,
and the temptation for micromills to enter this business for small market
areas, many existing minimills will have to close or to alter their product

mix. The spread of ladle metallurgy plus the improvements in steel making and rolling have opened up the possibility of producing "special bar quality" and low-alloy grade bars in existing minimills. These same developments are also encouraging new minimills to be built to produce high-grade alloy bars (for example, Timken Company) and even stainless bars. These new mills would compete with existing integrated and specialty producers.

New rolling techniques that have reduced the advantages of scale, better sorting of scrap to reduce tramp elements, and the combination of ladle metallurgy and better steel making and casting are all encouraging minimills to move into the production of structural shapes. For example, Chaparral, Bayou, and Northwestern Steel and Wire Company are now producing medium-sized structural shapes, and in 1986 Nucor announced a joint venture to produce even larger structural shapes. There is no reason why the minimills cannot move on to larger structural shapes, competing directly with the existing integrated producers. Similarly, minimills could produce rails, although precise metallurgy is essential. The real barrier to rail production may be the long-standing relationship between the three major integrated producers and the railroads. The advantages of minimill technology are evident, however, and either the major producers' market stranglehold will be challenged or they will reformulate their processes to take on minimill characteristics, with electric furnaces, casters, a rail mill, and minimill strategies.

Minimills have already begun to move into production of seamless tubes, where they have clear cost advantages. However, in the seamless tube business very precise product specifications are absolutely essential, particularly to the oil industry. Minimills, like their integrated rivals, must meet the American Petroleum Institute's specifications, but this will not be enough to convince a very careful market. It will be essential for minimills to develop market relationships before building seamless tube facilities. Integrating backward from a seamless pipe distributorship is one possibility. Another is acquiring a distribution network, perhaps from an integrated producer who wishes to abandon superfluous product lines. Given the size of minimills, a narrow product range must be maintained in production. It is possible to market many types and sizes of pipe without producing the full range by purchasing from other domestic or foreign producers. Welded tubes, especially of smaller diameters, are also likely to be produced in minimills (Interprovincial Steel and Pipe Corporation in Canada is already doing so). Minimills may be built for this purpose, or existing mills (like Lone Star Steel Company) may be rationalized and rebuilt to become minimill operations relying on scrap, producing a smaller product range, and utilizing simpler, more up-to-date technology.

The major hurdle for minimills is in sheet and plate products. Electric furnace mills (such as Phoenix Steel Corporation and United States Steel's Baytown plant) are currently producing plates. Such mills, however, are not noted for their efficiency or minimill-type organization. A plant under construction at Tuscaloosa, Alabama, is designed to produce 500,000 tons of coiled plate, initially from purchased slab, but such a plant could have its own electric furnaces and continuous caster.

The real breakthrough in flat products will come with the application of thin slab casting or mini hot-strip mills. Minimills utilizing these technologies are likely to be built before 1990. Nucor, in particular, began to enter this arena in 1986. The earliest facilities will target the market for commercial grades of plate and sheet, including those for barrels, pails, and structural applications. Higher-quality products will require further improvements in technology and may take some time.

Semifinished Products

Minimills already provide some semifinished steel to integrated producers in the form of billets, and this activity will expand as minimill quality improves with ladle metallurgy. However, even greater potential for minimills may lie in selling slabs to those firms with good rolling mills but inefficient iron and steel-making facilities. Such minimills would use electric furnaces and slab casters to convert scrap to regular, continuously cast slab for use in hot-strip (or plate) mills designed for eight-inch to twelve-inch slabs. There are many integrated producers who will be purchasing slabs, especially in and around Pittsburgh and Chicago. Trade restrictions that limit imports of these slabs will almost certainly create a requirement for slabs from minimills (as well as from the integrated firms' own steel furnaces and slab casters). Under normal market circumstances, these requirements are likely to be between 5 million and 10 million tons, and imports confined to less than 2 million tons if current trade restrictions remain. Minimills built for this purpose would have to secure their market through contractual obligations with the integrated rolling mills; but they might eventually take over the whole process and displace the integrated producers altogether. For several such mills, feasibility studies show production costs for slabs of less than $200 (in 1985 dollars) per ton. This is yet another example of how the minimills will continue to erode the integrated firms' position.

As usual, the minimill approach to entering or capturing new markets will be to single out a fairly low grade of a product category and specialize, using lower costs to push aside the integrated producer. Subsequently, the minimills extend into the integrated domain, learning about the market and improving their product quality as they go. This is done with mills of modest scale, each specializing in a very narrow product range, with some perhaps moving to higher product grades, but more likely with successive mills building on the gains of the previous mills and aiming for steadily higher market niches.

As table 6-2 shows, minimills may provide more than 25 percent of the industry's shipments by 1990 and 40 percent by 2000. The integrated mills will have to adopt strategies that are more market oriented, offering narrower product ranges, simplified up-to-date technology, lower input (particularly labor) prices, and a complete dedication to the control of production costs to hold their projected 42 percent of the U.S. market. This does not mean that all integrated producers will become scrap-based minimills, but that they must restructure their operations to lower their costs and develop lower-cost technologies to stay cost competitive in their remaining product lines.

The Emphasis on Distribution and Value Added

Traditionally the steel industry has emphasized volume of steel production, with success measured in tons produced. The current steel crisis has revealed the fallacy of this approach. Further, the market changes under way have fragmented customer requirements into ever more precise grades and types of steel, and into higher product qualities, limiting long production runs. In part this has benefited minimills, since market changes have partly offset scale advantages. At the same time, however, the requirement for higher quality and more precise metallurgy is forcing minimills to upgrade their products. Market changes have also radically increased the role of service centers, who purchase steel from the producers, warehouse it, and resell it, with or without such minor fabrication as slitting, shearing, drilling, and bending. These distributional organizations are also taking on the characteristics of Japanese trading companies, financing and marketing the products of others.

The role of service centers has increased dramatically in recent years. In 1980 they received approximately 19 percent of U.S. steel shipments and by 1985, 25 percent.[3] For imported steel, these organizations handle an even higher share. In flat products the growth of service centers has been spectacular, rising from about 23 percent of shipments in 1980 to almost 50 percent in 1985. This in part represents the integrated producers' abandonment of labor-intensive activities to nonunion firms; it also reflects the increased specialization by steel firms, which no longer act as a supermarket to their customers but supply a limited product range to a distributor who handles the products of a number of producers. The major service centers have been remarkably profitable, with average returns on sales of over 5 percent between 1983 and 1985.

The changes under way in the steel industry indicate that distributors and traders will take a larger and larger share of steel shipments— probably more than 50 percent of industry shipments by 1995. For now these organizations are, with some exceptions, relatively small operations. However, this too will change and major organizations will proliferate largely through merger, in some cases with a steel producer partner (for example, Joseph T. Ryerson and Son with Inland Steel). These organizations will continue to handle products of both minimills and integrated producers, although some distributors will deal mostly with minimill products. The role of these companies is also expected to expand in another way—into product fabrication. Some do minor fabrication and others are adding coating lines (painting, electro-galvanizing, and so on). This kind of activity will proliferate because of the advantage of nonunion labor rates and the ability of these distributors to provide specialized services for customers. Steel companies can be expected to pass on some of their finishing activities to these downstream organizations, and steel consumers, such as the automotive firms, may return fabrication activities (such as blanking) to them.

Distributors have the advantage of being more in tune with the market and more flexible than producers since they are not burdened with an unwieldy integrated apparatus and rigid union contracts. Such organizations could well come to dominate the steel business, creating a whole new layer of activities between steel firms and the final consumer of

3. AISI, *Annual Statistical Report, 1984,* table 15; *1985,* table 13.

products made from steel. This will be beneficial to all steel producers, and even more so to the producers who participate in ownership. With this step, the transition of the steel business from a resource orientation to a market orientation will be completed, within a span of less than twenty years.

The expanding role of distributors will have an impact on integrated and minimill steel producers alike, but principally on the large firms in the short run. Minimills will continue to upgrade their products, and as they do so, to avoid the risks of too much specialization, they must either work with a distributor or move on into steel fabrication such as wire drawing, mesh manufacture, fasteners, and so on. In a sense the steel industry is becoming more of a service industry, and this will delineate the future of both the integrated and nonintegrated components of the industry.

Implications for Public Policy

Steel is often cited as the prime example of U.S. deindustrialization. For one reason or another, the domestic steel industry is believed to have lost its competitiveness and to have settled into decline. Yet within this industry, a group of firms is rising phoenixlike from the ashes. It is likely that by the end of the century the United States will boast a highly competitive small-scale steel sector that accounts for nearly 40 percent of the country's steel requirements. This sector will provide a vital raw material for economic growth at prices competitive with the lowest-cost producers in developing countries.

Why has there been rejuvenation in this industry, but not in other smokestack industries such as copper, farm machinery, or heavy equipment? Obviously, it is not because demand for steel has been rekindled in the 1980s. In 1985, U.S. steel consumption was about 20 percent below its 1973–74 peak.[4] Nor is it because a national policy of trade protection has resuscitated dying firms. Rather, it is because small, entrepreneurial firms have relentlessly pressed electric furnace technology while their larger competitors continued to try to emulate the success of the Japanese in the 1960s and 1970s.

4. Ibid., *1985*, table 1A; *1982*, table 1A.

Comparison with Other Industries

The decline in Big Steel is not unlike that in a number of industries in which the United States has lost competitiveness. In cameras, electronic goods, and even automobiles, U.S. firms have often been slow to adapt to changes in technology or taste. American companies failed to apply the latest solid-state technology in radio and television design until long after the Japanese had wrested a large share of the world market in these products, and they were very slow to use computers in the manufacture of cameras.

Even in the automobile industry, where General Motors began to recognize the need for smaller cars before the 1973–74 oil crisis, the major producers find it difficult to compete with the Japanese in producing smaller cars. All U.S. companies lagged badly in installing robots. Product reliability declined, and "fit and finish" quality slipped badly relative to that of Japanese companies. American producers are planning to import large numbers of cars from Japan and even South Korea while pursuing joint ventures with Japanese companies to produce subcompact cars in the United States. At best, General Motors hopes to launch a competitive car by 1989–92—nearly twenty years after beginning to plan for smaller cars.

The difference between steel and these other industries is that new domestic firms with new managements emerged to challenge the established domestic steel firms and the major foreign producers. The large steel companies did not foresee the potential in small-scale operations. Most continued to build or modernize large-scale facilities even when demand slowed and scrap prices fell. When they built electric furnaces, they constructed large vessels that required long heat times and could not be easily integrated with continuous casters. The furnaces were seldom built with an efficient, small-scale operation in mind but rather as a low-cost alternative to what management preferred—the construction of blast furnaces and basic oxygen furnaces.

Similarly, the larger integrated firms could not undo the legacy of decades of labor-management relations that had guaranteed large real wage increases without equivalent increases in productivity. The new minimill competitors were not bound by these labor contracts, nor by the local work rules that accompanied them. As a result, the integrated firms found themselves saddled with much higher unit labor costs than the minimills for every product line in which the two competed. This labor-cost disadvantage could not be reduced without a complete change in the philosophy of organizing plants and, of course, in industrial relations. Although the integrated firms are now moving away from industrywide collective bargaining, even changes in their 1986 collective bargaining agreements with the United Steelworkers are unlikely to reduce their labor costs per ton to the level enjoyed by the minimills.

In the consumer electronics and automobile industries there were no new domestic entrants to try new technologies or offer new products. Perhaps this was because there is no analogue to the $75 million steel minimill plant. But in 1970 the minimill was hardly viewed as a potential competitor of United States Steel or Bethlehem Steel. Today, the giants are neither actual nor potential competitors of most of the new minimills, having closed most of their simpler, small-diameter finishing facilities. The entrepreneurs in the steel industry, such as Nucor's chairman, F. Kenneth Iverson, saw the potential in a technology and transformed it rapidly into a far more efficient production system than the traditional system comprising large blast furnaces and basic oxygen furnaces.

Comparison with Other Countries

The U.S. minimill sector is not unique. The Bresciani in Italy have thrived as very small electric-furnace mills, supplying simple bar products. There is a large minimill sector in Spain. Electric furnaces now account for almost 30 percent of steel production in non-Comecon countries (table 6-3), but in many cases these furnaces are in large, government-owned complexes, not in dynamic entrepreneurial firms.

Table 6-3 identifies the countries with the largest share of electric furnace production, which provides a rough approximation of the relative importance of their minimill sectors. In the European Community in 1983, only Italy and the United Kingdom had a sizable share of their steel capacity in small, independent steel firms, and even their shares were declining under the EC steel cartel's rationalization policies. Independent electric furnace companies exist in West Germany and Belgium, but their share of the market is much more modest than the minimill share in Italy or the United Kingdom. The reason for the limited growth of minimills in most EC countries is quite simple. During the period of the most rapid improvements in electric-furnace and continuous-billet-casting technology, the major integrated firms in Europe encountered severe economic difficulties, and their governments poured

Table 6-3. *Electric Furnace Share of Raw Steel Production in Noncommunist Countries, 1983*

Area and country	Percent
Total non-Comecon countries	29.09
European Community	25.86
Italy	53.48
United Kingdom	29.97
Other Western Europe	37.35
Spain	55.77
Sweden	51.24
Latin America	37.63
Argentina	55.20
Brazil	24.91
Mexico	45.04
North America	29.88
United States	30.44
Canada	26.62
Africa, Asia, and Oceania	26.94
South Korea	29.09
Taiwan	32.02
Japan	28.42
South Africa	26.90

Source: International Iron and Steel Institute, *Steel Statistical Yearbook 1984* (Brussels: IISI, 1984), p. 7.

billions of dollars of subsidies into the firms. At the same time, a succession of adjustment plans has sought to turn the EC steel sector into a cartel, forcing minimills to struggle (often unsuccessfully) against the politically more powerful nationalized integrated companies.[5] Thus, independent steel producers have found it difficult to obtain quotas and to compete with nationalized firms when they do.

In Latin America, there are many independent minimills in both Brazil and Argentina. And a large share of steel production in Venezuela and Mexico is from electric furnaces, but they are generally owned by large nationalized firms that use directly reduced ore as their raw material. In Asia, Japan, South Korea, and Taiwan have vigorous minimill sectors. Surprisingly, both Taiwan and South Korea have large, efficient nationalized integrated companies but appear to tolerate a dynamic minimill sector. In both cases, the nationalized firm concentrates on sheet and plate products while the independent companies supply the others. It

5. These were a series of "anticrisis" plans imposed by the European Coal and Steel Community beginning in the 1970s.

seems unlikely that these independent companies will be in the forefront of the rush toward thin slab casting and minimill sheet production.

In short, a number of foreign countries have dynamic minimill sectors, but in many cases government support of integrated companies leads to open discouragement of new, competitive firms. Obviously, the United States' reluctance to support its dying integrated firms has been an important component of the U.S. minimills' success.

Policies to Stimulate New Firms

The comparison of European and U.S. steel policies provides an excellent example of how government can help new firms by simply standing aside, refusing to subsidize failing firms and allowing new growth to replace the fallen timber. The new entrants in the minimill sector were neither encouraged nor encumbered by the U.S. government. In the 1970s, a few loan guarantees were offered to steel producers by the Economic Development Administration of the Department of Commerce, but these were not directed at integrated companies with wire rod or bar mills. As a result, minimills suffered no comparative disadvantage.

Trade protection, in the form of voluntary restraint agreements in 1969–74, may have helped the smaller firms since they induced the Japanese and Europeans to concentrate their exports on higher-valued products. And the trigger prices of 1978–82, established to limit imports, also helped somewhat by increasing the prices of steel imports. Of course, the latter policy may also have delayed the retirement of United States Steel's and Bethlehem Steel's wire rod facilities.

More important, the large, failing steel companies have not been subsidized in the United States as in Europe. As a consequence, integrated firms in the United States have not been given an unnatural advantage over minimills as they have in Europe, nor has the U.S. government been induced to turn the industry into a cartel and thereby exclude new entrants.

As long as U.S. minimills are able to compete for scrap, capital, electrical energy, and technology with other steel makers in the United States, it is likely that they will expand their market share. Their success will depend on the government's continued willingness to let failing firms fail and to be at least neutral toward new firms. If, however, the Lockheed or Chrysler bailouts should become the model for steel policy, the dynamic new competition in this industry could easily be thwarted by a government eager to defend its most recent mistake.

Trade Protection and the Supply of Capital

The most common defense of trade protection for a troubled industry is that it is the only mechanism available for generating the cash flows required for reinvestment in modernization of the industry. The industry generally argues that it is being starved of the needed capital to reinvest and become truly competitive. This starvation, of course, arises from poor returns on past investments and the perception that future investments are not likely to generate any better returns. If trade protection is designed to raise steel prices "temporarily" so that investment in integrated steel assets may be increased, it is flying in the face of a consensus in capital markets that such investments are not profitable.

The minimills have not benefited much from past episodes of trade protection. Nor have they complained about lack of capital. Since 1970 their production capacity has increased by nearly 15 million tons. At a current cost of approximately $300 per raw ton of capacity, this suggests a total investment of about $4.5 billion (at 1985 values) over fourteen years. In addition, most minimills have spent substantial amounts on modernization of furnaces, casters, and rolling mills during a period of extreme hardship and generally declining real investment in the integrated sector. Capital has come from Australian, Canadian, Japanese, French, and West German sources as well as U.S. capital markets. Private investors have underwritten a number of mills without resort to public equity issues. In short, there has been a substantial flow of capital because of the promise of returns on the minimill investments. Access to capital is rarely a problem when prospective profits loom. The U.S. integrated sector, in contrast, suffers from a shortage of capital in the sense that prospective returns do not justify many of its modernization plans.

The Right to Fail

Most minimill investments are made with relatively short investment horizons. Plants are designed to be modified in a few years to accommodate the ever-improving technology. Given their small size, however, none can anticipate a sympathetic government if it were to fail. Bankruptcies have occurred, and a number of reorganizations or failures are likely if the U.S. dollar rises substantially or the economy turns into recession. Even such relatively large minimills as Bayou might not be safe under this combination of economic events. One of the most important competitive spurs to constantly improving technology, manpower usage, and product quality is the fear of insolvency. Inefficient minimills could not survive very long in the highly competitive bar and rod markets in which minimills operate. As long as there is no government guarantee that a minimill will survive in the face of economic uncertainty, these companies will continue to press aggressively against the technological frontier of electric-furnace steel making.

Conclusion

The steel industry in the United States has suffered from negative real growth for twelve years. Imports have risen as steel consumption has declined. Moreover, there is little prospect for a reversal of these trends. Despite this gloomy backdrop, new entrants are appearing daily in the form of minimills. These newer companies have grown substantially in the face of declining steel demand, and they are likely to continue to grow over the rest of this century. In short, small companies are arising to take market share away from the steel giants despite weak demand and depressed U.S. steel prices.

The minimills' success clearly is due in part to low scrap prices, but it is also the result of their efficient and flexible management. They have been better able than the integrated firms to adapt to new technology and even to push out the frontiers of this technology. They have moved progressively into product lines previously thought to be closed to them for technical reasons. They have avoided rigid union contracts and inflexible work rules and, as a result, have enjoyed far better labor productivity than their larger, integrated rivals.

The U.S. experience has not been repeated in most developed countries. In part, this may be due to higher scrap prices elsewhere, but it is also due to governmental policy. In European countries, minimills have not replaced integrated firms because the European Community has chosen to turn its integrated firms into a cartel, requiring government quota allotments for each ton produced. Potential minimill entrants find it very difficult to obtain quota allotments. Smaller electric furnace companies have been more successful in Spain (before its entry into the EC), in Japan, and in South Korea. These countries have not discouraged minimills in the name of rationalizing their integrated sectors.

In the United States, government steel policy has consisted largely of various forms of import control. Import restrictions have raised U.S. steel prices, further reducing consumption of steel from its already depressed level. However, the restrictive import policies have not inhibited the expansion of the minimill sector. In fact, the higher domestic prices of steel products have probably served to accelerate the entry of these newer entrepreneurial companies. In turn, this new entry and declining consumption growth have accelerated the rate at which the large integrated companies have had to close their noncompetitive facilities. Most of the closures have been concentrated in the smaller products that the minimills currently produce, not the larger sheets and plates.

By the end of this century, the U.S. steel industry will be somewhat smaller, but the larger, integrated companies will be substantially smaller. These integrated companies have already shed 30 percent of their mid-1970s capacity. And they are likely to lose nearly 50 percent of their current capacity by the end of the century. During the same period, the more dynamic new minimills will double their current output and capacity. In short, the complexion of the steel industry will have changed dramatically by the end of the century. No longer will U.S. steel simply be Big Steel.

Tables

Chapters 2, 3, and 4 contain numerous estimates of minimill and integrated producers' costs in the United States and other countries. This appendix provides greater detail for many of these estimates, which were drawn from extensive interviews with numerous steel makers and from proprietary data made available to the authors.

Each table contains costs for each major process in steel making—coking, smelting (blast furnaces), refining (basic oxygen furnaces and electric furnaces), casting, and rolling. The cumulative costs through each stage are also shown. Given the variance in costs from one plant to another, these estimates should be regarded as representative costs in 1985, not the precise costs of any single mill or plant.

Table A-1. *Cost of Producing Wire Rod at a Representative Steel Minimill Operating at 90 Percent of Capacity, 1985*

Item	Input per ton of output	Operating cost (dollars per ton)
Electric furnace[a]		
Scrap	1.10 tons @ $85.00	93.50
Labor	0.90 man-hour @ $17.50	15.75
Electricity	520 kilowatt hours @ $0.045	23.40
Natural gas	0.27 thousand cubic feet @ $4.50	1.20
Fluxes and alloys	. . .	8.00
Other additives	. . .	3.50
Electrodes	9 pounds @ $1.20	10.80
Refractories, for electric furnaces	22 pounds @ $0.25	5.50
Refractories, for ladles and tundishes	16 pounds @ $0.25	4.00
Supplies	. . .	9.00
Miscellaneous	. . .	4.70
Total raw steel cost		179.35
Rod mill		
Yield loss, from raw steel	0.060 ton @ $179.35	10.75
Scrap credit	0.045 ton @ $85.00	−3.85
Labor	0.85 man-hour @ $17.50	14.90
Electricity	160 kilowatt hours @ $0.045	7.20
Natural gas	2.25 thousand cubic feet @ $4.50	10.15
Supplies	. . .	6.50
Miscellaneous	. . .	4.00
Total rod mill cost		49.65
Sales and overhead		
Labor, shipping	0.35 man-hour @ $17.50	6.15
Labor, administration	0.25 man-hour @ $17.50	4.40
Miscellaneous	. . .	4.50
Total sales and overhead		15.05
Financial costs		
Depreciation	. . .	9.00
Interest	. . .	12.00
Taxes	. . .	2.00
Total financial cost		23.00
Total cost of finished steel		267.05
Addendum		
Man-hours per ton = 2.40[b]		

Source: Authors' calculations based on data provided by various producers. Tons are net tons in all tables unless otherwise noted. Costs are rounded.

a. Continuous casting of billets.

b. Includes all labor, operating and administrative.

Table A-2. *Cost of Producing Wire Rod at a Low-Cost Steel Minimill Operating at 90 Percent of Capacity, 1985*

Item	Input per ton of output	Operating cost (dollars per ton)
Electric furnace[a]		
Scrap	1.09 tons @ $85.00	92.65
Labor	0.80 man-hour @ $17.50	14.00
Electricity	485 kilowatt hours @ $0.045	21.85
Natural gas	0.50 thousand cubic feet @ $4.50	2.25
Fluxes and alloys	. . .	8.00
Other additives	. . .	5.50
Electrodes	7 pounds @ $1.20	8.40
Refractories, for electric furnaces	20 pounds @ $0.25	5.00
Refractories, for ladles and tundishes	14 pounds @ $0.25	3.50
Supplies	. . .	8.00
Miscellaneous	. . .	2.00
Total raw steel cost		171.15
Rod mill		
Yield loss, from raw steel	0.040 ton @ $171.15	6.85
Scrap credit	0.028 ton @ $85.00	−2.40
Labor	0.70 man-hour @ $17.50	12.25
Electricity	150 kilowatt hours @ $0.045	6.75
Natural gas	2.05 thousand cubic feet @ $4.50	9.25
Supplies	. . .	7.00
Miscellaneous	. . .	2.00
Total rod mill cost		41.70
Sales and overhead		
Labor, shipping	0.20 man-hour @ $17.50	3.50
Labor, administration	0.20 man-hour @ $17.50	3.50
Miscellaneous	. . .	4.00
Total sales and overhead		11.00
Financial costs		
Depreciation	. . .	12.00
Interest	. . .	18.00
Taxes	. . .	2.00
Total financial cost		32.00
Total cost of finished steel		255.85

Addendum
Man-hours per ton = 1.95[b]

Source: Data provided by producers. Costs are rounded.
a. Continuous casting of billets.
b. Includes all labor, operating and administrative.

Table A-3. *Cost of Producing Wire Rod at a New Steel Minimill Operating at 90 Percent of Capacity, 1985*

Item	Input per ton of output	Operating cost (dollars per ton)
Electric furnace[a]		
Scrap	1.08 tons @ $85.00	91.80
Labor	0.55 man-hour @ $17.50	9.65
Electricity	435 kilowatt hours @ $0.045	19.60
Natural gas	0.50 thousand cubic feet @ $4.50	2.25
Flux and alloys	. . .	8.00
Other additives	. . .	5.50
Electrodes	5 pounds @ $1.20	6.00
Refractories, for electric furnaces	18 pounds @ $0.25	4.50
Refractories, for ladles and tundishes	12 pounds @ $0.25	3.00
Supplies	. . .	8.00
Miscellaneous	. . .	1.45
Total raw steel cost		159.75
Rod mill		
Yield loss, from raw steel	0.035 ton @ $159.75	5.60
Scrap credit	0.025 ton @ $85.00	−2.15
Labor	0.60 man-hour @ $17.50	10.50
Electricity	140 kilowatt hours @ $0.045	6.30
Natural gas	1.90 thousand cubic feet @ $4.50	8.55
Supplies	. . .	8.00
Miscellaneous	. . .	1.50
Total rod mill cost		38.30
Sales and overhead		
Labor, shipping	0.20 man-hour @ $17.50	3.50
Labor, administration	0.15 man-hour @ $17.50	2.65
Miscellaneous	. . .	4.00
Total sales and overhead		10.15
Financial costs		
Depreciation	. . .	22.00
Interest	. . .	16.50
Taxes	. . .	2.00
Total financial cost		40.50
Total cost of finished steel		248.70

Addendum
Man-hours per ton = 1.55[b]

Source: Data provided by producers. Costs are rounded.
a. Continuous casting of billets.
b. Includes all labor, operating and administrative.

Table A-4. *Cost of Producing Cold-Rolled Coil at a Representative Integrated Steel Mill Operating at 90 Percent of Capacity, 1985*

Item	Input per ton of output	Operating cost (dollars per ton)
Coke oven		
Coal	1.56 tons @ $55.00	85.80
Labor	0.85 man-hour @ $22.50	19.15
Electricity	30 kilowatt hours @ $0.045	1.35
Other energy	4.10 million Btus @ $4.50	18.45
By-product credit	9.00 million Btus @ $4.50	−40.50
Miscellaneous	. . .	12.15
Total coke cost		96.40
Blast furnace		
Iron ore	1.53 tons @ $40.00	61.20
Coke	0.53 ton @ $96.40	51.10
Labor	0.80 man-hour @ $22.50	18.05
Electricity	35 kilowatt hours @ $0.045	1.60
Other energy	2.6 million Btus @ $4.50	11.70
By-product credit	4.0 million Btus @ $4.50	−18.00
Reserve, for relining	. . .	4.75
Miscellaneous[a]	. . .	9.05
Total hot metal cost		139.40
Basic oxygen furnace		
Hot metal	0.82 ton @ $139.40	114.30
Scrap	0.29 ton @ $80.00	23.20
Labor	0.45 man-hour @ $22.50	10.15
Electricity	60 kilowatt hours @ $0.045	2.70
Fluxes and additives	. . .	11.75
Other additives	. . .	5.00
Supplies	. . .	7.00
Miscellaneous	. . .	2.55
Total liquid steel cost		176.65
Slab caster		
Yield loss, from raw steel	0.025 ton @ $176.65	4.40
Scrap credit	0.018 ton @ $80.00	−1.45
Labor	0.45 man-hour @ $22.50	10.15
Electricity	30 kilowatt hours @ $0.045	1.35
Refractories	. . .	5.00
Miscellaneous	. . .	6.00
Total slab caster cost		25.45
Ingot caster		
Yield loss, from raw steel	0.030 ton @ $176.65	5.30
Scrap credit	0.021 ton @ $80.00	−1.70
Labor	0.20 man-hour @ $22.50	4.50
Miscellaneous	. . .	7.50
Total ingot caster cost		15.60

Table A-4 (continued)

Item	Input per ton of output	Operating cost (dollars per ton)
Slab mill		
Yield loss, from ingots	0.15 ton @ $192.25	28.85
Scrap credit	0.105 ton @ $80.00	−8.40
Labor	0.45 man-hour @ $22.50	10.15
Electricity	25 kilowatt hours @ $0.045	1.15
Other energy	1.10 million Btus @ $4.50	4.95
Miscellaneous	. . .	5.65
Total slab mill cost		42.35
Total average slab cost[b]	. . .	223.25
Hot-strip mill		
Yield loss, from slab	0.050 ton @ $223.25	11.15
Scrap credit	0.035 ton @ $80.00	−2.80
Labor	0.65 man-hour @ $22.50	14.65
Electricity	110 kilowatt hours @ $0.045	4.95
Other energy	2.15 thousand cubic feet @ $4.50	9.70
Miscellaneous	. . .	12.80
Subtotal		50.50
Total hot-rolled coil cost		273.75
Cold-finishing process[c]		
Yield loss, from hot band	0.120 ton @ $273.75	32.85
Scrap credit	0.084 ton @ $80.00	−6.70
Labor	1.63 man-hour @ $22.50	36.70
Electricity	170 kilowatt hours @ $0.045	7.65
Other energy	1.80 thousand cubic feet @ $4.50	8.10
Miscellaneous	. . .	22.00
Subtotal		100.60
Total cold-rolled coil cost		374.35
Sales and overhead		
Labor, shipping	0.30 man-hour @ $22.50	6.75
Labor, administration	0.45 man-hour @ $22.50	10.15
Miscellaneous	. . .	11.00
Total sales and overhead		27.90
Financial costs		
Depreciation	. . .	24.00
Interest	. . .	12.00
Taxes	. . .	7.20
Total financial cost		43.20
Total cost of finished steel		445.45
Addendum		
Man-hours per ton = 5.75[d]		

Sources: Authors' calculations based on confidential data.
a. Includes sintering costs; assumes 25 percent of iron ore is sintered.
b. Assumes 35 percent slab casting and 65 percent rolled ingots at a total average slab cost of $223.75.
c. Pickling, tempering, annealing, and cold rolling.
d. Includes all labor, operating and administrative.

Table A-5. *Cost of Producing Cold-Rolled Coil at a New Integrated Steel Mill Operating at 90 Percent of Capacity, 1985*

Item	Input per ton of output	Operating cost (dollars per ton)
Coke oven		
Coal	1.56 tons @ $55.00	85.80
Labor	0.65 man-hour @ $22.50	14.65
Electricity	25 kilowatt hours @ $0.045	1.15
Other energy	3.65 million Btus @ $4.50	16.45
By-product credit	9.00 million Btus @ $4.50	− 40.50
Miscellaneous	. . .	14.50
Total coke cost		92.05
Blast furnace[a]		
Iron ore	1.51 tons @ $40.00	60.40
Coke	0.48 ton @ $92.05	44.20
Labor	0.55 man-hour @ $22.50	12.40
Electricity	30 kilowatt hours @ $0.045	1.35
Other energy	2.4 million Btus @ $4.50	10.80
By-product credit	4.2 million Btus @ $4.50	− 18.90
Reserve, for relining	. . .	4.00
Miscellaneous	. . .	11.00
Total hot metal cost		125.25
Basic oxygen furnace		
Hot metal	0.82 ton @ $125.25	102.70
Scrap	0.29 ton @ $80.00	23.20
Labor	0.40 man-hour @ $22.50	9.00
Electricity	50 kilowatt hours @ $0.045	2.25
Fluxes and additives	. . .	7.00
Other additives	. . .	9.00
Miscellaneous	. . .	10.60
Total liquid steel cost		163.75
Slab caster		
Yield loss, from raw steel	0.020 ton @ $163.75	3.30
Scrap credit	0.014 ton @ $80.00	− 1.10
Labor	0.30 man-hour @ $22.50	6.75
Electricity	25 kilowatt hours @ $0.045	1.15
Refractories	. . .	4.75
Miscellaneous	. . .	6.65
Subtotal		21.50
Total slab cost		185.25

Table A-5 *(continued)*

Item	Input per ton of output	Operating cost (dollars per ton)
Hot-strip mill		
Yield loss, from slab	0.040 ton @ $185.25	7.40
Scrap credit	0.028 ton @ $80.00	−2.25
Labor	0.50 man-hour @ $22.50	11.25
Electricity	100 kilowatt hours @ $0.045	4.50
Other energy	2.00 thousand cubic feet @ $4.50	9.00
Supplies	. . .	6.50
Miscellaneous	. . .	6.50
Subtotal		42.90
Total hot-rolled coil cost		228.15
Cold-finishing process		
Yield loss, from hot band	0.050 ton @ $228.15	11.40
Scrap credit	0.035 ton @ $80.00	−2.80
Labor	1.00 man-hour @ $22.50	22.50
Electricity	120 kilowatt hours @ $0.045	5.40
Other energy	1.10 thousand cubic feet @ $4.50	4.95
Miscellaneous	. . .	23.50
Subtotal		64.95
Total cold-rolled coil cost		293.10
Sales and overhead		
Labor, shipping	0.30 man-hour @ $22.50	6.75
Labor, administration	0.40 man-hour @ $22.50	9.00
Miscellaneous	. . .	10.00
Total sales and overhead		25.75
Financial costs		
Depreciation	. . .	94.60
Interest	. . .	71.10
Taxes	. . .	10.50
Total financial cost		176.20
Total cost of finished steel		495.05

Addendum
Man-hours per ton = 3.85[c]

Source: Data provided by producers.
a. Includes sintering costs; assumes 25 percent of iron ore is sintered.
b. Pickling, tempering, annealing, and cold rolling.
c. Includes all labor, operating and administrative.

Table A-6. *Cost of Producing Cold-Rolled Coil at a New Steel Minimill Operating at 90 Percent of Capacity, 1985*

Item	Input per ton of output	Operating cost (dollars per ton)
Electric furnace		
Scrap	1.08 tons @ $85.00	91.80
Labor	0.45 man-hour @ $17.50	7.90
Electricity	450 kilowatt hours @ $0.045	20.25
Natural gas	0.50 thousand cubic feet @ $4.50	2.25
Fluxes and additives	. . .	11.50
Other additives	. . .	2.00
Electrodes	6.0 pounds @ $1.20	7.20
Refractories	. . .	5.00
Miscellaneous	. . .	7.50
Total raw steel cost		155.40
Slab caster		
Yield loss, from raw steel	0.025 ton @ $155.40	3.90
Scrap credit	0.018 ton @ $85.00	−1.55
Labor	0.35 man-hour @ $17.50	6.15
Electricity	25 kilowatt hours @ $0.045	1.15
Refractories	. . .	5.00
Miscellaneous	. . .	7.00
Subtotal		21.65
Total slab cost		177.05
Hot-strip mill		
Yield loss, from slab	0.050 ton @ $177.05	8.85
Scrap credit	0.035 ton @ $85.00	−3.00
Labor	0.55 man-hour @ $17.50	9.65
Electricity	115 kilowatt hours @ $0.045	5.20
Natural gas	1.90 thousand cubic feet @ $4.50	8.55
Miscellaneous	. . .	11.75
Subtotal		41.00
Total hot-rolled coil cost		218.05
Cold-finishing process[a]		
Yield loss, from hot band	0.065 ton @ $218.05	14.15
Scrap credit	0.045 ton @ $85.00	−3.85
Labor	1.45 man-hours @ $17.50	25.40
Electricity	160 kilowatt hours @ $0.045	7.20
Natural gas	1.60 thousand cubic feet @ $4.50	7.20
Miscellaneous	. . .	19.60
Subtotal		69.70
Total cold-rolled coil cost		287.75
Sales and overhead		
Labor, shipping	0.35 man-hour @ $17.50	6.15
Labor, administration	0.40 man-hour @ $17.50	7.00
Miscellaneous	. . .	10.00
Total sales and overhead		23.15

Table A-6 *(continued)*

Item	Input per ton of output	Operating cost (dollars per ton)
Financial costs		
Depreciation	. . .	46.95
Interest	. . .	35.20
Taxes	. . .	4.95
Total financial cost		87.10
Total cost of finished steel		398.00
Addendum		
Man-hours per ton = 3.70[b]		

Source: Data provided by producers.
a. Pickling, tempering, annealing, and cold rolling.
b. Includes all labor, operating and administrative.

Table A-7. *Cost of Producing Concrete Reinforcing and Special Quality Bars and Seamless Tube at a Representative Steel Minimill Operating at 90 Percent of Capacity, 1985*

Item	Concrete reinforcing bars	Special quality bars	Seamless tube[a]
Operating costs	225.00	260.00	461.00
Labor	31.50	52.50	80.00
Scrap	93.00	95.00	106.00
Energy	40.00	42.00	52.00
Miscellaneous	60.50	70.50	223.00
Financial costs	8.50	26.00	179.00
Depreciation	4.00	10.00	100.00
Interest	3.50	14.00	70.00
Taxes	1.00	2.00	9.00
Total costs	233.50	286.00	640.00

Source: Data provided by producers.
a. Projected costs for tube 5.5 inches in diameter and 0.275 inch thick.

A Simulation Exercise: Scrap Availability for Electric Furnaces

The supply of scrap depends on a number of variables: steel-making yields from raw steel to finished product, the yield loss in steel fabricating, the return rate of obsolete scrap, and the mix of furnaces. This supply (S) may be represented as:

$$(1) \qquad S = [R_i Y_i Y_{fi} + 1 - Y_i Y_{fi}](OH + BOF)$$

$$+ [R_m Y_m Y_{fm} + 1 - Y_m Y_{fm}] ELEC$$

$$+ [R_l Y_{fl} + (1 - Y_{fl})]I,$$

where R_i, R_m, and R_l are the return rates for obsolete scrap from steel products supplied by integrated firms, minimill companies, and imports, respectively; Y_{fi}, Y_{fm}, and Y_{fl} are the yields in steel fabricating from integrated, minimill, and imported steel shipments; Y_i and Y_m are the steel-making yields in integrated and minimill steel works, respectively; OH, BOF, and $ELEC$ are the total quantities of raw steel produced by open hearths, basic oxygen furnaces, and electric furnaces, respectively; and I is steel imports. (Scrap imports are ignored.)

Equation 1 abstracts from reality in two important respects. First, it is essentially a steady-state model in which the recovery rate for obsolete scrap is a constant fraction of current steel consumption. In fact, obsolete scrap is recovered from steel products made over a period of twenty years or more. If steel consumption is constant over time, however, obsolete scrap recovery may be written as a function of *current* steel consumption. Second, electric furnaces at integrated works are included in the minimill side of the supply equation. This assumes that the yields at minimills and integrated firms in products produced from electric furnaces are the same. While the assumption is dubious, it is not likely to make a major difference in any illustrative calculations.

The demand for scrap is likely to be a rather simple function of steel making at each type of furnace plus export demand. The share of scrap charged into basic oxygen furnaces, open hearths, electric furnaces, and blast furnaces may vary with relative prices, but for purposes of illustration the rates are assumed to be constant. The demand for scrap (D) then is simply:

$$(2) \qquad D = 0.52\,OH + 0.27\,BOF + 1.1\,ELEC$$
$$+ 1.1\,FOUN + ED,$$

where $FOUN$ is foundry output and ED is export demand. The ratio of scrap to raw steel production in open hearths and basic oxygen furnaces includes some charge of scrap into the blast furnaces that produce the hot metal charge.

If reasonable values are assigned to the variables in equation 1, the balance of scrap supply and demand can be reduced to a very simple equation. Assume that the obsolete return rate (R_i, R_m, and R_l) is 0.50; integrated and minimill yields (Y_i and Y_m) are 0.70 and 0.85, respectively; and fabrication yields (Y_{fi}, Y_{fm}, and Y_{fl}) are 0.85. If export demand is about 10 percent of total scrap supply, then supply for domestic purposes may be calculated as:

$$(3) \quad S = [0.9]\{[(0.50)\,(0.70)\,(0.85) + 1 - (0.85)\,(0.70)]\,[OH + BOF]$$
$$+ [(0.50)\,(0.85)\,(0.85) + 1 - (0.85)\,(0.85)]\,[ELEC]$$
$$+ [(0.50)\,(0.85) + (1 - 0.85)]\,I\}.$$

Setting equation 3 equal to equation 2 less export demand (ED) and dividing by total raw steel ($RAWSTEEL$), the equilibrium of S and D can be simplified to:

$$(4) \quad 0.633\,\frac{OH+BOF}{RAWSTEEL} + 0.575\,\frac{ELEC}{RAWSTEEL} + 0.518\,\frac{I}{RAWSTEEL}$$
$$= 0.27\,\frac{BOF}{RAWSTEEL} + 0.52\,\frac{OH}{RAWSTEEL}$$
$$+ 1.1\,\frac{ELEC}{RAWSTEEL} + 1.1\,\frac{FOUN}{RAWSTEEL}.$$

As a check on these parameters, insert the current values for the various furnace shares and the import share of steel consumption. In the first half of the 1980s, imports have been about 21 percent of apparent consumption or about 20 percent of U.S. raw steel production at current yields. Production in basic oxygen furnaces has been about 60 percent of U.S. raw steel production, in open hearths about 9 percent, and in electric furnaces about 31 percent.[1] Foundry output is equal to an average of 0.15 times raw steel. The left-hand side of equation 4 at these values is equal to 0.708 while the right-hand side is equal to 0.709.

1. American Iron and Steel Institute, *Annual Statistical Report, 1984* (Washington, D.C.: AISI, 1985), tables 1A, 1B.

A Model of the U.S. Iron and Steel Scrap Market

A model of the scrap market should include equations for export demand, domestic demand, the supply of home scrap, the supply of prompt industrial scrap, and the supply of obsolete scrap. Because there are no data that distinguish obsolete from prompt industrial scrap, a model of the market must be drawn from only two scrap supply functions, one for all purchased scrap (*PS*) (including obsolete and prompt industrial as well as scrap destined for exports) and one for home scrap (*HS*):

$$(1) \qquad\qquad PS_t = a_0 + a_1 FABR_t + a_2 [e^{a_3 TIME_t}] P_t$$

and

$$(2) \qquad HS_t = b_0 + b_1 RAWSTEEL_t + b_2 FOUN_t + b_3 CONCAST_t,$$

where *FABR* is sales of steel mill products to domestic users (including net imports), *RAWSTEEL* is the total production of raw steel, *FOUN* is foundry output, *CONCAST* is the amount of raw steel that is continuously cast, *P* is the real price of scrap, *t* is the current period, and *TIME* is a time trend. The price term *P* includes the possibility of growth in the price responsiveness of scrap supply over time because of the improvement in scrap collection technology and the increasing accumulation of obsolete scrap. Not all obsolete scrap is collected, and the remainder does not oxidize as rapidly as it accumulates; hence, there is the possibility that improving technology increases the recycling of discards from previous years. *CONCAST* in equation 2 allows for a declining trend of home scrap supply due to better yield performance in steel mills caused by the introduction of continuous casting.

It is conceivable that there are lags in the operation of the supply side of the market, but it is by no means clear that such lags are important. Annual data are used to estimate the model, and because these lags appear to be less than a year, equation 1 utilizes only current values of scrap price and industrial activity.

Scrap is bought by both domestic and foreign buyers, so the analysis of demand is separated into domestic and foreign components. The domestic demand function for scrap should contain variables for steel production, foundry production, the price of scrap, and the price of pig iron (the principal substitute for scrap in steel making). The domestic demand function can be specified either by taking the output of steel produced in each type of furnace as given, or by assuming that the mix of steel produced in open hearths and electric and basic oxygen furnaces varies with the relative prices of scrap and pig iron. Since the latter formulation (in which steel output is exogenous) should capture the effect of relative scrap prices on the choice of furnace, domestic demand can be represented as:

$$(3) \qquad DD_t = c_0 + c_1 RAWSTEEL_t + c_2 FOUN_t$$
$$+ c_3 (P_t/PIGP_t).$$

The variable $P_t/PIGP_t$ is the ratio of scrap price to the price of pig iron. As scrap prices fall relative to pig iron prices in a recession, the production of steel is shifted from basic oxygen to electric furnaces and even to open hearths—to the degree that plant structure permits.

Foreign demand for U.S. exports of scrap (FD) is assumed to be a function of foreign steel production, the value of the dollar, the real price of scrap, and a time trend to capture the shift toward electric furnaces in other countries:

$$(4) \quad FD_t = d_0 + d_1 FORSTEEL_t + d_2 DOLLAR_t + d_3 P_t + d_4 TIME_t,$$

where FORSTEEL is non-U.S. world production (excluding communist countries) of raw steel and DOLLAR is a trade-weighted index of the value of the dollar.

The market-clearing equilibrium condition is simply that:

$$(5) \qquad\qquad PS_t + HS_t = DD_t + FD_t.$$

Inventory swings are accommodated by including only *consumption* of scrap in DD and deducting inventory accumulations at steel mills from total market supply. Thus, $PS_t + HS_t$ is the current period's supply from all sources, including home scrap, less inventory accumulation.

The empirical specification of the model is based on annual observations since it appears that market adjustments in supply and demand occur within a year's time. The model was estimated over the period 1961–83 and over more recent periods to ascertain whether there have been significant structural shifts over time. Given the recent growth in

Table C-1. *Mean Values of Variables in an Econometric Model of the Scrap Market, 1961–83*

Variable	Mean value	Standard deviation
	Millions of tons[a]	
DD = U.S. scrap consumption[b]	85.2	12.8
FD = U.S. net scrap exports[b]	7.8	1.8
PS = purchased scrap (including net exports)[b]	44.6	7.8
HS = home scrap[b]	48.0	8.2
$RAWSTEEL$ = raw steel production[c]	122.4	18.5
$FABR$ = steel consumption[c]	97.8	14.6
$FOUN$ = foundry shipments[b]	16.2	2.8
$CONCAST$ = steel produced by continuous casting[c]	10.0	9.4
$FORSTEEL$ = foreign steel production[d]	522.6	130.9
	Index	
$DOLLAR$ = trade-weighted value of U.S. dollar[e]	111.0	11.8
P = composite real price of scrap (deflated by GNP deflator)[f]	125.0	33.7
	Dollars per ton	
$PIGP$ = real price of pig iron (deflated by GNP deflator)[g]	43.46	10.20

a. Tons are net tons.
b. Data from U.S. Bureau of Mines.
c. Data from American Iron and Steel Institute.
d. Data from International Iron and Steel Institute.
e. Values from Federal Reserve Board. March 1973 = 100.
f. U.S. Bureau of Labor Statistics. Base year 1967 = 100.
g. *World Steel Dynamics;* U.S. Federal Trade Commission. In 1967 dollars.

electric furnaces and the growing importance of steel imports and scrap exports since the late 1960s, some such changes are possible, but they do not appear to be important. The mean values of the variables for 1961–83 are given in table C-1.

The econometric results of a two-stage least-squares estimate of equations 1–4 are as follows (*t*-statistics are in parentheses):

$$(6) \quad PS = -5.79 + 11.08\,Pe^{0.01\ TIME} + 0.356\,FABR - 7.81\,D74$$
$$\qquad\quad (2.19) \quad (8.71) \qquad\qquad (12.40) \qquad\quad (2.76)$$

$$\bar{R}^2 = 0.931; \rho = -0.341; \text{Durbin-Watson} = 2.07$$

$$(7) \quad HS = 1.98 + 0.082P + 0.280\,RAWSTEEL$$
$$\qquad\quad (1.06) \quad (0.09) \qquad (5.42)$$
$$\qquad - 0.178\,CONCAST + 0.824\,FOUN$$
$$\qquad\quad (4.13) \qquad\qquad\quad (2.31)$$

$$\bar{R}^2 = 0.984; \rho = 0.400; \text{Durbin-Watson} = 1.58$$

(8) DD = 6.25 − 1.08 $P/PIGP$ + 0.456 $RAWSTEEL$
          (1.64)   (1.15)   (6.90)
        + 1.60 $FOUN$ + 7.12 $D74$
          (3.46)   (3.85)
          $\bar{R}^2$ = 0.986; ρ = 0.912; Durbin-Watson = 1.76

(9) FD = 4.81 − 0.070P − 33.0 $DOLLAR$
          (0.57)   (0.05)   (0.66)
        + 0.23 $FORSTEEL$ − 0.406 $TIME$
          (2.07)   (2.08)
          $\bar{R}^2$ = 0.275; ρ = 0; Durbin-Watson = 2.053

The estimates of the supply relationships indicate that the supply of home scrap, equation 7, is not significantly related to the price of scrap. Operating practices are simply not very sensitive to annual variations in the price of scrap. However, the supply of purchased scrap (including net exports of scrap) is significantly related to price—alternative trials suggest that this price sensitivity is growing at a rate of 1 percent per year. These estimates suggest price elasticities of 0.36 at the point of means, or 0.39 by 1983 at similar mean price and quantity. For obsolete scrap alone, which is about half of total purchased scrap, this translates into an elasticity of about 0.80.

The estimated home scrap supply function is unremarkable. The yield from raw steel to finished product in steel mills is, implicitly, 72 percent without continuous casting and 90 percent with continuous casting. Both estimates are very close to yields realized in current practice.

The domestic demand for scrap appears to have a small negative price elasticity, but this elasticity is only 0.04 at the point of means.

Finally, export demand is only mildly sensitive to the value of the dollar and is not significantly affected by domestic scrap prices. When the period is shortened to 1961–79 or 1961–80, the value of the dollar assumes a statistically significant coefficient, but when the equation is extended into the 1980s, the significance of the value of the dollar declines dramatically. Surprisingly, there is evidence of a downward trend in export demand, perhaps because foreign sources of scrap are becoming more important.

Index

Acs, Zoltan, J., 68n
Adams, Walter, 36n
American Iron and Steel Institute, 12, 43
Argentina, minimills, 110
Armco, 49, 53, 55
Australia, iron ore, 32, 43, 44
Automobile industry, 2, 108

Baldwin, C. Y., 69n
Barnett, Donald F., 6n, 32n, 36n
Bars. *See* Steel products
Basic oxygen furnace (BOF): operating
   costs, 30, 32; production capacity, 5, 45;
   replacement of open hearths by, 3, 36,
   38; use of scrap, 74, 77–78, 126–27
Bayou Steel Corporation, 10, 33, 103, 113
Bethlehem Steel Corporation, 3, 38, 43n,
   47, 53, 55, 83, 100, 111
Big Steel. *See* Integrated steel companies
Billets. *See* Steel products
Blast furnace, 3, 39
Blooms. *See* Steel products
BOF. *See* Basic oxygen furnace
Brazil, 12, 38; iron ore, 32, 43, 44; mini-
   mills, 26, 110

California Steel and Tube, 6, 7
Canada, 84
Capital: costs, 25, 27, 45; international
   sources, 69–70, 112; minimill access to,
   68–69, 112
Carron, Andrew S., 87n, 93
Cartels, European Community steel, 109,
   110, 113
Chaparral Steel Company, 10, 14, 23, 33,
   103
Clark, J. P., 69n
Coal prices, 45, 85–86, 87
Cold-rolled steel coil, 52–53, 66–67
Competition, in steel industry: foreign, 2,
   26–27, 32, 35, 36, 97, 100; integrated
   plant versus minimill, 19, 34, 39–40;
   among minimills, 16
Congressional Research Service (CRS),
   89, 91, 94
Continuous casting process, 5, 6, 19, 52,
   58, 72

Cooke, Henry, 3n, 6n, 69n
Copperweld Steel Company, 10, 69
Cordero, Raymond, 3n, 6n, 69n
Costs, steel production: attempts to re-
   duce, 54–55; and exchange rates, 29, 46–
   47; in integrated plants versus minimills,
   1, 2, 25–28, 40, 65–67; and prices, 32–
   34; U.S. versus foreign, 45–47
Cotton, Richard, 90n
Crandall, Robert W., 36n
CRS. *See* Congressional Research Service

Directly reduced iron (DRI), 73, 84, 97
Dirlam, Joel B., 36n
Dollar, U.S. *See* Exchange rates
DRI. *See* Directly reduced iron

Economic Development Administration,
   111
Economies of scale, minimill, 19, 25
Edison Electric Institute, 93
Efficiency, production: BOF, 3, 5; minimill, 1, 9, 20, 23, 27, 28
Electric furnace, in integrated plants, 3,
   18, 52, 53, 74–75; in minimills, 5, 7, 9,
   18–19, 20; operating costs, 30, 84–85;
   technological improvements in, 18, 28,
   56–58, 60–61; use of scrap, 73, 78–79,
   84, 126
Electricity: costs, 85–86; future supply and
   demand for, 88–93; industrial rates, 86–
   87, 93–94
Employment, U.S. steel industry, 100
Energy costs, 85–94
Energy, Department of, 86, 94
European Community, 12, 13; DRI consumption, 84; scrap availability, 99; steel
   cartels, 109, 100, 113
Exchange rates: and steel production costs,
   29, 45–47; and U.S. steel market, 13, 19,
   29; and world steel production, 12, 13
Exports, U.S. steel, 13, 74

Flat-rolled products. *See* Steel products,
   sheets and plates
Florida Steel Corporation, 1, 14, 18, 23
France, 12

133

Up from the Ashes